CU00706585

Telecommunications in Transition

Telecommunications in Transition

Michel Carpentier

Director General, Telecommunications, Information Industries and Innovation,
European Communities Commission DG XIII

Sylviane Farnoux-Toporkoff

Professor, Université Paris VIII
Doctor of Economic Sciences

C Garric

Advisor to the Telecommunications Directorate DG XIII,
European Communities Commission

Translated from the French by **C.P. Skrimshire**, UK

JOHN WILEY & SONS
Chichester · New York · Brisbane · Toronto · Singapore

Copyright © 1992 Commission of the European Communities

Published in 1992 by John Wiley & Sons Ltd.
Baffins Lane, Chichester
West Sussex PO19 1UD, England

Other Wiley Editorial Offices

John Wiley & Sons, Inc., 605 Third Avenue,
New York, NY 10158-0012, USA

Jacaranda Wiley Ltd, G.P.O. Box 859, Brisbane,
Queensland 4001, Australia

John Wiley & Sons (Canada) Ltd, 22 Worcester Road,
Rexdale, Ontario M9W 1LI, Canada

John Wiley & Sons (SEA) Pte Ltd, 37 Jalan Pemimpin   05–04,
Block B, Union Industrial Building, Singapore 2057

British Library Cataloguing in Publication Data

A catalogue record for this book is available from the British Library
ISBN 0 471 93190 X

Typeset in 10pt Bookman by Text Processing Dept, John Wiley & Sons Ltd, Chichester
Printed and bound by Biddles Ltd, Guildford and King's Lynn

Contents

Foreword   vii

Acknowledgments   xv

1   The Seismic Revolution in America   1

    Introduction   1

   1.1   The history of American telecommunications   2

   1.2   The American telecommunications market in 1983
       on the eve of the deregulation of the sector   7

   1.3   The reappraisal of the structure of the
       telecommunications market   16

   1.4   The shock of the breaking up of AT&T   19

   1.5   The operation of the American market following
       deregulation and the breaking up of AT&T   24

   1.6   A permanent regulation   31

   1.7   Progress of American regulation after the break up
       of AT&T   36

   1.8   Today's services and equipment market in the
       United States. Perspectives on the 1993 horizon   39

   1.9   Overall assessment   44

2   The First Steps Towards a Community Policy   47

   2.1   The premises for community action in the information
       technology domain   48

   2.2   The first community actions on telecommunications
       matters   53

   2.3   Standardization in information and telecommunications
       technologies   64

**3   The Rise in Power of European
Telecommunications**   73

   3.1   European telecommunications in 1984—an assessment. . . .   73

   3.2   The first shock waves of American deregulation   85

3.3 The achievement of the single market 87

3.4 1984–1992: towards a global European
telecommunications policy 90

4   Towards New Industrial Equilibria 145

4.1 Assessment 145

4.2 In conclusion 160

4.3 Epilogue 164

List of Acronyms 169

Index 173

Foreword

Telecommunications is at the forefront of current economic and social affairs. The digitalization of networks has destroyed the boundaries which separated them from the fields of electronics and computing. The resultant multiplication of tele-services and their increasing sophistication have exploded the technical and commercial limitations which previously existed. This evolution has been decisive and has followed from technical facts, and not from political will. Humanity, viewed as a global organism, is witnessing the installation, growth and diversification of an extended nervous system equipped with prodigious efficiency. The pressure for technical innovation is oblivious of the geography of nations; it challenges cultural identities.

The torrent of change carries everything before it. State monopolies, regulations, and the interests of powerful equipment manufacturing and operating companies are all called into question. These events appear spectacular if one recalls the stability of the structures existing up to the end of the 1970s in a protected, cocooned, crystallized, century-old world of telephony in which firms, guaranteed of their incomes, flourished.

This book is the story of this global jolt. The authors have tried to show how the shifting of the American tectonic plate in this field has propagated towards our continent, and how the European Community and its member states have reacted to the changes.

The first part of the work is dedicated to this American "earthquake". The most spectacular developments came about under the Reagan administration, one of whose main characteristics in European perception was the relinquishing of the power of the state in favour of private economic forces. What came out of this? The American Telegraph and Telephone Company, AT&T, was the most powerful company in the world by virtue of its longevity, its traditions, its almost total control of the immense telecommunications market in the USA, by the sheer size of its workforce and shareholder base, by the weight of its industrial interests, and by its unequalled scientific research and technical potential. We saw this great monopoly power being obliged to enter into a dialogue in which it bargained for its destiny and ultimately agreed to its own dismantlement. Who were its inquisitors? A small but powerful independent commission, the FCC, a representative of the Department of Justice "DOJ", and Judge Harold Greene.

The FCC (Federal Communication Commission) was created in 1934 by a provision of the Communication Act. From that time onwards, the parliamentary representation of American government put the regulation of the telecommunications domain and the operating rules of the private players into the hands of this Commission. Federal Judge Harold Greene, who was to play a considerable role throughout the long duration of the most revolutionary phase of this story, acted in the name of the anti-trust laws, in complete independence of FCC and the political process. His one key concept was the necessity to create opportunities for competition in this field for the benefit of users.

In stages, by amicable consent ("consent decree"), the process would result in the break up of AT&T into several concerns, new definitions of territory available to the new players, and the opening of the market to competitors who had hitherto been kept in the wings by the effects of a monopoly.

In fact, in examining the effects of the measures approved by the FCC and Judge Greene, we did not find a situation of open competition at all levels, but rather one of new regulation. It is therefore erroneous to speak merely of "deregulation". If some believe naively that American liberalization has been equivalent to "laissez faire-ism" in an unregulated environment, then I urge them to pay careful attention to the account of this adventure. It is worth putting the following question: How many private European industrial groups would be ready to embark on such a process, and to end up with an amicable compromise of such a magnitude without protesting against the abuse of power by the government?

It is necessary to unravel the American regulatory mechanisms, from which much can be learned: politicians have put organizational bodies in place and have delegated to them various powers. These bodies have then sought by dialogue an acceptable regulatory formula with private firms. The market has been, to a large extent, continuously accompanied by regulatory procedures applied by organizations protected from day to day political contingencies.

The European Community is rightly taking pains to find supranational solutions which do not excessively undermine the authority and responsibility of national governments. It would seem to me that the community would gain by drawing inspiration from this American democratic experience, but nothing is straightforward in today's world. It will be necessary to continue to monitor the moves and countermoves enacted during the course of the formal processes which shape regulatory power in the United States.

The enormity of the issues has naturally provoked challenges and may lead to a re-examination of the distribution of roles amongst the elements of government, the FCC, the DOJ, and the Court. The process is still continuing. We are witnessing an evolving situation. The regulators modify the attitudes of the players who react, via pressure groups, upon the regulators themselves.

In summary, we find ourselves, thanks to the American telecommunications adventure, operating in a social and economic laboratory which is rich in lessons to be learned about what we should know for the future operation of a "controlled liberal system".

Personally, I have much admiration for how our American colleagues have recognized what needed to be done. The measures adopted have generated internal competition, where it was necessary, as an instrument to stimulate initiatives in the marketplace. However, in practice, control over national interests has been built into fundamental industrial and business operations. At the same time, the openness in principle to international competition has rendered the policy, in legal terms, virtually unassailable. This elegant result has been obtained from collaboration between intelligent partners thanks to the smooth running of the "recursive dialogue" process, to use the phrasing of Edgar Morin.

How has the EC, and particularly the Communities Commission, reacted to these same technical developments which impact on the very foundations of the nervous system of our continent? An informed account of this is presented in detail in this text.

Telecommunications cannot be separated into the network part, that is the technologies which make up the "container", and the actual information services which form the "contents". The measures taken by the Communities Commission have been applied to the whole arena considered as a single entity. This stance is all the more remarkable since, at a national level, these factors have for a long time been, and continue to be, treated as separate variables.

The reader will not fail to be struck by three observations which arise from the concise account of community actions given by the authors:

—community initiatives did not begin until after 1972;

—pressure from the Commission was virtually the sole driving force;
—operations were very modest to begin with. They did not amount
to more than palliative doses until after 1983.

The initiatives did not enter their preparatory phases until 1972,
by which time decisive moves were already being made by the
Americans (and then subsequently the Japanese), using powerful
resources. The European governments were content to operate in
isolation by promoting their own national champions, and only
deployed sub-critical resources. Three fields were then conquered by
our main competitors: microtechnology, administration infrastructure
computing, and to a lesser extent domestic electronics. The external
trade statistics did not reveal the scale of the damage until ten years
later, but the battles on these fronts were lost even before Europe had
the notion of profiting from its community dimension. In computing
technology, leaving aside the special case of military electronics, the
only domain genuinely preserved was that of telecommunications,
by virtue of the protection from predation that had been established.
This fact should never be forgotten.

From 1973 onwards, the initiatives were the result of the
Commission acting on its own rather than of pressure from member
governments. This was evidenced by the inertia that existed in
making political decisions on proposals of action, however modest,
which came from the Commission. There is a cultural explanation for
this imbalance. The Commission, through its newly appointed administrators and through the personalities called on to form the consultative committees, was forewarned of the high stakes associated with
the advent of the "communication civilization". It was aware of the
decisive character of the effects of scale in the coming battles of
innovation.

The national governments were less receptive. They were more
alert to short term preoccupations, and their policies were still more
attuned to the "energy civilization" which corresponded to their traditional industries. In any case, governments which had backed
their national champions were now loath to watch them be under-
mined by community agreements.

The scene changed in the period 1981/3, but it was once more
due to initiatives of the Commission, notably through the skill of
Vice President Etienne Davignon who concluded an alliance, in principle, with the main leaders of the information technology industry.
Telecommunications was not yet directly concerned, but the
launching of the Esprit programme pointed the way.

From this time onwards, the deliberations in Brussels also served as a model "recursive dialog" laboratory. The problems were dealt with by three groups of partners; the Commission, which provided the frame of reference of the work, the representatives of the monopolist national operators, and the industrialists. From out of this consultation emerged a will for concerted action which gave rise to "the Green Paper" on telecommunications. The process was exemplary in the sense that it catalyzed the realization of a common destiny among players who had previously sought to maintain their distance and differences. Exemplary, also, because it opened the way to operational programs having a supercritical dimension after reaching an understanding on a common doctrine concerning these objectives.

The Green Paper and the Resolution of June 1989 left it to the member states to decide for themselves the organizational forms which best suited their traditions and particular structural circumstances, and which would best facilitate the functioning of their networks, while setting in motion the mechanisms that would assure the separation of the regulatory functions from those of commercial exploitation. It was understood that the Community, for its part, would use its best offices, delegated to it under the Treaty, to formalize the rules and principles of competition which would apply in this field, and to harmonize the regulations concerning the free availability of equipment and services. This experience has been of great importance. In order to appreciate the path taken to reach this stage, it is necessary to go back to the circumstances which surrounded the initial studies and consultations. In 1983, national monopolies, eager to preserve their autonomy, were still on the scene. Certainly, they were used to meeting in international forums, and they did not wait for the Communities Commission to act on all issues before reaching certain agreements, especially in the area of international technical standardization. They were aware that they represented significant power bases whose roots had been deeply established in the past. In all member countries, the PTTs were in some respects states within states, protected by their technical nature, barricaded in many cases behind their own specifications and particular methods of standardization, supported by well organized unions and by collaborating industrial groups. Over the years, a symbiotic web had been woven between the involved parties which had given them a mutual feeling of comfort and security. This communal existence, at the heart of the PTT monopolies had in addition very significant peripheral social extensions. The jobs involved were very numerous. The employees claimed to have a strong sense of public duty; wrongly as some think, but correctly in the minds of others.

The quality of service concerned not only the PTT fraternity, but also impinged on the lives of businesses and the public, due to the fundamental role of communications and its impact on people and institutions. And then along came an external body, a General Directorate of the Commission, which, before the Treaty was even modified so as to establish the "single market", proposed to level the ground in the name of European homogeneity and to take into account the fears that the effects of scale could introduce. The haste was unexplained when everything in the past had been achieved through different mechanisms, primarily the policy of national purchasing preferences.

That the overall response, in free and open debate, has finally been positive is encouraging to all those who believe in the need to build a Europe on a real community basis. The result has been an agreement of general scope having enough force to guarantee rules ensuring sufficient harmonization, as well as recognition of the margins of freedom of all parties to take into account the diversity of their situations. This evolutionary process towards consent seems to typify the approach necessary in order to achieve a politically united Europe. We are no longer in the weakening climate of the hierarchical pyramid, which is led from the top, and in which the members obey. We now use a new approach which responds to the imperatives of freedom, and technical complexity, as well as efficiency.

The final part of this book will stimulate those who wish to have a view into the future and to understand the opportunities and vulnerabilities of the telecommunications scene in Europe, particularly the struggle against underutilization of resources, business equilibrium, and technological independence. What is above all striking is the fluidity of the scene. Industrial reorganizations are multiplying. New market positions are attained through alliances, mergers, takeovers, and pre-emptive technical and business strategies. A communal Europe is far from being in total charge of its own ground, and is penetrated by subsidiary companies of American and Japanese multinationals. This globalization undoubtedly has benefits, and its progress is welcome, but not at any price. It is necessary to question the vision of a Europe open to all directions, operating on its own a truly open market in the global theatre, but with reciprocity going by the board. In the key areas of information technology, telecommunications constitutes the only sphere where, through its historical heritage, it retains powerful players, freedom of choice, scope for initiatives, and regulatory systems. We cannot lose these trump cards without counterbalances. Freedom of the capacity for action is not only important for Europeans in the community. The eastern block countries are requesting aid to regenerate their economies.

Telecommunications equipment figures among the keys to development; Western Europe should be in charge of all its decisions in this sphere. Saying that is not to advocate following a model of a European nationalism whose deviations might correctly be feared. It is simply to advise steering towards balanced relationships, in a reciprocity of interactions between the three members of the triad: North America, Western Europe and Japan.

The work of Michel Carpentier, Sylviane Farnoux-Toporkoff and Christian Garric provides an objective description. The authors tell the story; they give an account of situations as they may be seen by an impartial observer. I believe that the setting out of these facts needs to be complemented by additional reflection on the part of senior politicians, technocrats, and union representatives and also of consumer organizations. How should we proceed further with the success story of the European telecommunications arena which underpins the transformation of our civilization and the international distribution of work?

The authors have bestowed a great honour on me in inviting me to write this foreword. As it happens, it is an honour which doubles with pleasure and recognition. It is indeed a pleasure to again encounter Sylviane Farnoux-Toporkoff and to have enjoyed her confidence in several consultations during the preparation of the work. I knew her as a little girl at the time of the signing of the Treaty of Rome when, with her father, we contributed to building the European electronics industry. I did not expect to find her, thirty years later, as a well informed economist specializing in technical, legal and political matters in the field of telecommunications, and enriched by the influential circle of personal contacts she has managed to establish with the best specialists in Europe and the United States.

My sentiments of recognition go to those in authority in the European Communities Commission who have risen to this challenge. This is not a personal tribute, but a tribute in the name of all those Europeans involved in the success of information and telecommunications technologies, the researchers, engineers, technicians, and employees in the specialist administration and service industries. A handful of men in Brussels and Luxembourg have realized the importance to the community of the technological revolution which was affecting electronics, computing, telecommunications, information technology and language-based industries. From this realization emerged a determination.

The awakening of governments and industrialists has been fed by studies and recommendations for action. Little by little, momentum for action emerged, mediated by agreements and planned programs.

I have had the opportunity to participate in this effort. I can there-fore testify to the intelligence, the tenacity and the enthusiasm of the Commission teams. Since 1985 they have been reinforced by the political vision and the remarkable success attained by the Commission presided over so brilliantly by Jacques Delors along the route to European construction. These teams were not working for themselves. Had they simply let the time pass, no one would have accused them of inaction. They have struggled for the science, tech-nology, industry and the independence of the community.

In casting my mind back to earlier years, I can only recall some names and must commit the injustice of omitting many, but I take pleasure in mentioning the role played by Christopher Layton, direc-tor from 1973 to 1981; by Christian Garric, whose extreme modesty cannot hide the soundness of his judgement, the balance of his ideas and his unerring competence; by Gunter Schuster, former Director General of DG XII; by Roland Hueber, who continues today to develop a strategic imagination worthy of much praise; by Jean Marie Cadiou, who launched the Esprit program in such a remarkable way. Herbert Ungerer should also be cited for the essential role that he played in the preparation of the Green Papers on telecommunica-tions and satellites. I have already acknowledged the decisive political importance of the initiatives of Etienne Davignon. Michel Carpentier, through his untiring work, has capitalized on his qualities of initiative, originality and conciliation, has allowed them their full rein, and has pursued them with vigour, since, with the confidence and support of the Vice Presidents Davignon, Narjes, and Pandolfi, he has taken in hand the general running of the overall activities of the Commission.

It is appropriate that this book has been written by Michel Carpentier, who is known through his role as chief monitor and as head of this impressive series, and through the success of the pro-grams which are still running. We are only at the start of the road. All the forces are in play, but nothing is won. I hope that in five years' time a new work on this subject will appear. I feel sure it will reveal many surprises to us. The history of information technology and tele-communications is only just beginning.

André Danzin
International consultant
Council member of the Club of Rome

Acknowledgments

The authors wish to express their thanks to Harvey Seifter for his considerable assistance in preparing this translation for publication.

They would also like to express their thanks to John Arcale, President of Complain Associates, Inc. for his substantial contribution.

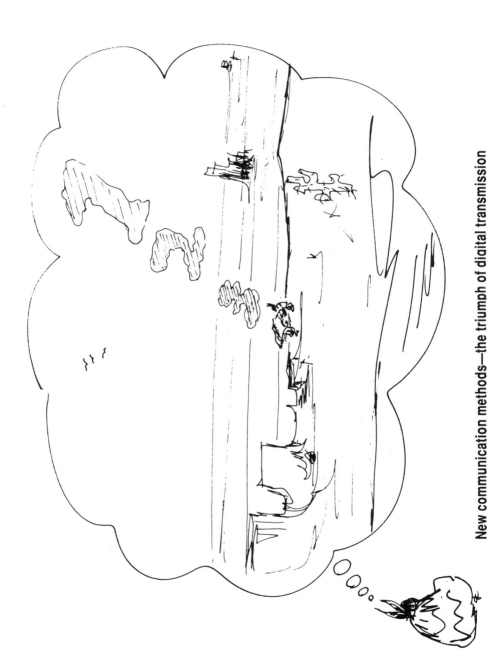

New communication methods—the triumph of digital transmission

1

The Seismic Revolution in America

INTRODUCTION

The American telecommunications scene (1) has been shaken since 1982 by a true revolution which has had a significant impact on every household and business in the country, and whose effects have been universally felt.

On January 1 1984, the American Telephone and Telegraph company (AT&T), for more than 100 years practically synonymous with the American telecommunications industry, and having a *de facto* monopoly of telecommunications in the United States, was broken up following a judicial decision. At the time, AT&T was the largest company in the world as measured by its assets (155 billion dollars) as well as by the number of its employees (990 000).

The *de facto* monopoly which AT&T had enjoyed was the result of the Communications Act passed by Congress in 1934 which had defined, and put in place, the rules governing telecommunications in the United States. With this Act, Congress responded to a consumer demand for legislation ending a confused state of affairs marked by ineffective competition, which had led to multiple uncoordinated services provided by over 10 000 companies, often duplicating each other, and using incompatible equipment.

---

(1) The function of the telecommunications services sector is to provide an information transmission service from one point to another, by means of apparatus arranged in networks (local and long distance) which are composed of three main equipment types.

- switching equipment, installed at the network nodes which link the terminals together and allow them to communicate with each other; in modern terms this means "exchanges".
- transmission equipment which carries the signals.
- terminal equipment connected at the extremities of the network (telephones, telex machines etc).

1.1   THE HISTORY OF AMERICAN TELECOMMUNICATIONS

The story begins with the telegraph transmission networks, set up in this country thanks, among others, to Morse in 1838, and which allowed the electric telegraph to become widely used.

By 1850, one could identify approximately twenty fiercely competitive companies. Soon, only the two most competitive, American Telegraph and Western Union survived; these companies amalgamated in 1866 and formed a single company, which from then on held a monopoly in this field: The Western Union.

1.1.1   The establishment of the monopoly

● *1876-1894: 18 years of exploitation of Graham Bell's patent by AT&T: the first monopoly*
The telephone appeared when Graham Bell (1) demonstrated voice transmission for the first time in Boston in 1876.

He then created the Bell Telephone Company which, thanks to the monopoly which his patent conferred on him for 18 years (that is up to 1894), and to the extraordinary momentum given it by Theodore Vail (2) whom he placed in charge, became the company which would be known from then on as "the Bell System".

In 1881, the latter company bought from Western Union its subsidiary, Western Electric, whose factory was considered to be the best in the United States. In 1885, Theodore Vail created a subsidiary given the responsibility of providing intercity and international links: The American Telephone and Telegraph company (AT&T).

● *1894: The patent comes into the public domain*
In 1894, Graham Bell's main patent came into the public domain, and all companies wishing to make use of it were able to do so freely: this marked the opening of competition, which, for the Bell Company, quickly led to a reduction in its market share and provoked it to restructure itself. In 1899, AT&T, formerly a subsidiary, found itself the group's parent. The name of the group was AT&T from then on.

In 1907 Theodore Vail was recalled to be the head of the company and, due to his effectiveness, AT&T rapidly regained its domination of the market. With a slogan which declared: "one policy, one system, one universal service", the notion arose of a single telecom-

---

(1) Graham Bell had the good fortune to lodge his patent on the instrument that was to be known as the "telephone" a few hours before Elisha Gray. In 1876, it was Bell's patent which was recognized as the only valid one and it therefore benefitted, according to American industrial property law, from an 18 year protection.
(2) Theodore Vail ran the company during two separate periods; the first between 1878 and 1887, then a second from 1907 to 1919.

munications network operated by a single company, in order to best satisfy the needs of the nation.

However, the competition in the market was such that, by 1910, over 10 000 companies were operating in this field in the United States.

In many areas, several of these companies co-existed. Often their equipment was not compatible with that of their competitors (for example, in Philadelphia, a customer wishing to be connected to the police, the fire brigade, and the hospitals had to obtain three different telephones which were available from several different companies).

Faced with this disarray resulting from multiple uncoordinated, incompatible, and redundant services, different States started to regulate the telephone companies as "public service" companies from about the turn of the century. By 1920, many states substituted public regulation for free competition in the telephone industry, chiefly through the enactment of two laws.

—The Mann-Elkins Act of 1910 which gave the Interstate Commerce Commission (ICC) control over the tariffs fixed by the telegraph, telephone and cable companies, regarded from then onwards as "Public Utilities" (1) in an attempt to rationalize the industry.

—The Public Transportation Act of 1920, which confirmed the authority of the ICC over the transmission of information (by cable or otherwise).

● *1921: The monopoly returns*
Because of the essential importance of telecommunications recently highlighted by the First World War, the inclination towards controlling the market was inevitable. Congress debated whether it should maintain free competition (2), nationalize the companies completely, as in other countries, or merely subject them to strict control. A consensus emerged in favour of the third option.

The two questions which then arose were:

—Which manufacturers should be permitted to remain active in this field?
—What prices should be fixed (these needing to be agreed in advance)?

These questions were addressed in 1921 by the Willis Graham Act and in 1934 by the Communication Act.

The Willis Graham Act (3), passed by Congress in 1921, gave the ICC the power to approve mergers between telephone companies (4),

---

(1) Public service organizations.
(2) In accordance with the principles of the Sherman Act of 1890.
(3) The advocates of this law considered that the telephone was a "natural monopoly".
(4) At the time there were 7 950 independent telephone companies.

in spite of the anti-trust laws, and in effect placed telecommunications under the control of the American authorities.

● *1934: The Communication Act*
It was above all the Communication Act of 1934 which actually established the rules which were to govern this field until the 1980s.

This law specified the rights and duties of "common carriers" (telecommunication network operating companies) conforming to a notion known as "universal service", which compelled them to put at the disposal of the American public (that is, without discrimination) to the extent possible, a rapid and effective cable or radio communications service covering the whole country and connected to the outside world, at a reasonable price.

This same Act created the FCC (Federal Communications Commission) in order to invest the Federal Government with control and regulatory authority overseeing the entire telecommunications sector.

The effect of the Act (and the concurrent advances in technology) was to reduce local (exchange) services to a single provider regulated by the states. The FCC only regulated interstate (the national network and the local access) services.

Faced with these requirements from the public authorities, only a single national (long distance) network run by a large company benefitting from significant economies of scale seemed capable of having a sufficiently efficient organization to be able to offer users the best service at the best price. The principal company able to claim conformance to the above was "of course" AT&T.

At the same time, a large number of mergers took place among the small non-Bell (independent) companies—more than 1000 of them survived and were permitted to provide local exchange services.

After this process had duly set out a path for telecommunications which seemed acceptable to all parties, the Congress did not intervene in a practical way until 1956, leaving the FCC to regulate telecommunications in step with the new technologies as they appeared: television from 1935, Telex (1945), satellites (1963), data (1978), and broad-band services (1983).

During this period, AT&T, through its control over the long distance network, was able to exercise effective control over the local companies and even to force them to sell out.

At the time of World War I, AT&T stopped buying exchange companies. This led to the growth of the large non-Bell companies (GTE, United, Alltel, Centel, Contel, etc.). These companies even cooperated with AT&T in the national (long distance) network.

However, the ambitions of the independent companies led them to contest the quasi-monopoly of AT&T. A large number of small companies merged amongst themselves. Their political influence, with both Congress and the FCC, enabled them little by little to contest the power of AT&T, which they judged to be excessive, and also to erode its commercial position each time a new technology arrived bringing new services and particularly buoyant markets. This phenomenon developed only slowly.

1.1.2   The first stirrings of the spirit of deregulation

From 1949, the Department of Justice "DOJ" criticized AT&T for having extended its telephone communications monopoly to the manufacture of telephone equipment by vertical integration in contravention of the Sherman Act.

The issue was resolved in 1956 by a "Consent Decree", in which AT&T undertook not to extend its sphere of activity into the computing sector (at the time the computing market was fairly small compared to telecommunications). The leading force in this field, IBM, was not put under any particular restriction. IBM, however, would not venture into the telecommunications sector before the 1980s. In compensation, AT&T was given the right to preserve its vertically integrated structure.

AT&T was also obliged to make available all of its patents to its competitors: existing patents free of charge, and future patents in exchange for reasonable fees.

Very quickly, numerous initiatives were taken by big and small manufacturers alike, and an extraordinary explosion of semiconductor applications was seen. It is therefore possible to trace the birth of the "Route 128" and "Silicon Valley" industrial development largely to the effects of the "Consent Decree", in whose absence the pioneers would have found themselves straight-jacketed by Bell patents and competition.

After this "Consent Decree", a number of legal actions were filed against AT&T. These were the first breaches of its monopoly since they allowed the emergence of competition. The following are probably the most significant:

● *The "Hush-a-Phone" decision of 1957*
AT&T prohibited its subscribers (under the pretext of operational efficiency) from using equipment not supplied by itself. "Hush-a-phone", which sold an instrument intended to guarantee confidentiality of calls using AT&T handsets, contested this prohibition. The

initial decision from the FCC was in AT&T's favour, but, upon review, the Court of Appeals authorized the free use of such instruments by subscribers.

● *The "above 890" decision of 1959*
AT&T also prohibited its subscribers from establishing for their own purposes point-to-point radio links using frequency bands above 890 MHz (which explains the origin of the name of the "above 890" decision). It forced users to lease the frequencies from it. In 1959, the FCC obliged AT&T to change its policy in this matter and, in 1960, even authorized users to employ their own point-to-point communication equipment, bypassing the traditional AT&T telephone network.

This decision was important, because it showed that technological evolution and the shift towards a service economy made the pursuit of a traditional regulatory policy more difficult.

● *The "Carterfone" decision of 1968*
AT&T denied subscribers the right to connect a receiver–amplifier to their telephone handset.

In 1968, the FCC found in favour of the Carterfone company in allowing the connection 'bf additional terminal equipment, not supplied by the telephone company, onto the telephone network provided that it contained equipment protecting the network of the telephone company.

In 1976, this measure was extended to include the provision of the main telephone in an. installation (except those belonging to AT&T, which were not to be covered until 1982).

In 1977 finally, the need to incorporate network protection equipment was removed, that is, direct connection was now possible as soon as the instrument concerned had been approved by the FCC. This was the opening of a whole new market: that of "interconnects".

● *The "MCI-specialized common carriers" decision of 1969*
With this decision, the FCC allowed competition in the supply of specialized point-to-point links, following an action by MCI. The FCC reasoned that such competition, having no harmful technical or economic effects on the operation of the telephone network, could only be beneficial to the user.

Other decisions were to intervene in opening new spheres of competition, notably that of satellites after 1971.

● *The "share and resale" decision of 1976*
The FCC allowed the renting out by AT&T at "high-volume" tariffs of telephone circuits to smaller operators who "sub-let" them to users, thus creating a new market; that of resale carriers.

● *The "execunet" decision of 1978*
When MCI started to compete with AT&T by providing a switched intercity service, AT&T accused them of moving into this market without the expressed agreement of the FCC. The latter found in favour of AT&T.

This decision was subsequently annulled by the Court of Appeals which sent the question back to the FCC. This time the Commission did not take action to protect AT&T and thereby allowed MCI and other companies (Other Common Carriers—OCCs), to offer the same switched intercity service as AT&T. Competition had been established in the public-switched network domain.

However, despite these legal actions, the telecommunications market in 1983 was still concentrated in the hands of AT&T, which played a quasi-monopolistic role in local-service and long-distance telecommunications, and in the manufacture of equipment through its subsidiary Western Electric. AT&T also held a dominant world position in research, the "Bell Laboratories", which developed a strategy deliberately oriented towards the future.

1.2 THE AMERICAN TELECOMMUNICATIONS MARKET IN 1983 ON THE EVE OF THE DEREGULATION OF THE SECTOR

1.2.1 AT&T (American Telephone and Telegraph)

The following account demonstrates both the simplicity and the complexity of the organization of the American market in 1983 and, specifically that of the AT&T group.

In 1983, AT&T was the premier company in the world, with 990 000 employees, and 155 billion dollars in assets. Its structure is represented in figure 1.

Its activities were divided into five domains, in each one of which it was dominant.

● *The operating services in the local networks* were devolved to 22 subsidiaries, called "Bell Operating Companies" (BOCs), each of

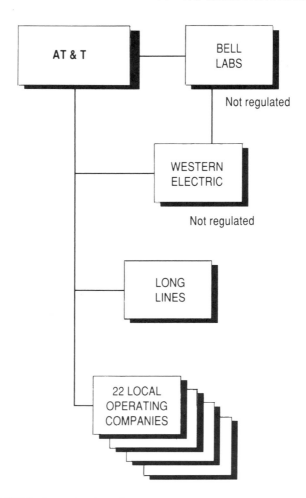

Figure 1   AT&T before the break up.

which enjoyed a geographic monopoly (represented in figure 2) and which, although only covering 60% of the national territory, handled 80% of the local traffic.

Each state was theoretically covered by a BOC. However, one BOC, Ohio Bell telephone (now part of Ameritech), served the state of Ohio, except Cincinatti and its suburbs.

Another company "Cincinatti and Suburban" (now Cincinatti Telephone) was not wholly owned by AT&T (only 20% owned), but was nevertheless part of the Bell system by contract.

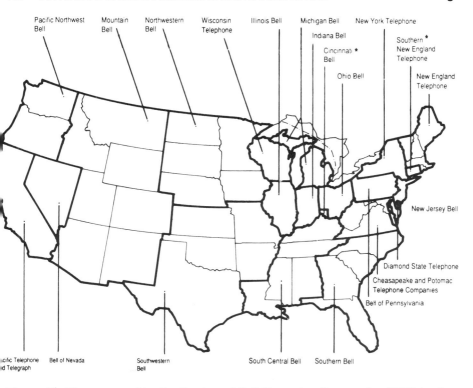

Figure 2 The geographic distribution of Bell Operating Companies (BOCs) prior to 1984. *:AT&T retained a minority holding in these two companies until 1984.

Some states included the serving territories of more than one BOC (e.g. Idaho).

● *The service operation of intercity networks* was covered by the "Long-lines" division which carried 96% of the long distance traffic (with a turnover of some 30 billion dollars) and which acted as common carrier for long distances on behalf of the 22 BOCs plus Cincinatti and SNET. Similarly, the state of Connecticut was served by Southern New England Telephone—part of the Bell System by contract—these two companies are not regarded as "independents", but as "Bell companies".

● *Western Electric,* a separate subsidiary, manufactured for the whole group the majority of the telecommunications equipment (including wires and cable) that it used. It sold 90% of its production internally (within the group) and made up 2/3 of the manufacturing activity in the American telecommunication industry. Its turnover was approximately 13 billion dollars.

● *In order to underpin its whole system, AT&T maintained the largest private research organization in the world: the Bell Telephone Laboratories,* where 7 Nobel Prize winners have won their fame, and where more than 20 000 patents have been lodged since 1925, including those relating to the transistor, the solar cell, stereophonic recording, and the UNIX operating system, which is at the heart of numerous computers.

For more than half a century, the "Bell Labs" have been regarded by scientists and information technologists as a model of organization and inventive efficiency, and have been one of the most remarkable forces of scientific, technical and industrial innovation in the United States.

● *The BOCS ran other activities,* such as the extremely profitable Yellow Pages publishing enterprise (business sections of the telephone directory).

● *However despite its domination of the internal American market, AT&T faced active competition in telecommunications services at the operational level, and in the equipment market.*

1.2.2 What competition did AT&T face?

● *In the area of telecommunications services operation, AT&T was confronted with three types of competition:*

—*Other local service providers in separate franchized territories*   The independent telephone operating companies provided local services (like the BOCs) and shared 20% of the subscribers (about 33 million telephones). They covered about 40% of the American territory (figure 3). They provided 4% of long distance telecommunications, so that their networks were completely interconnected with those of AT&T and the BOCs, the whole system forming a unified national telecommunications network.

The bulk of their activities revolved around providing services to rural communities and small towns. The zones that they covered represented a wide variety of climatic, geographic and demographic conditions.

■■■■   Independent telephone companies

☐   AT&T before deregulation

Source: *Behind the Telephone Debates* C.L. Weinhauss and A.G. Oettinger

Figure 3   Geographic distribution of the "independents" in 1983.

These enterprises ranged in size from small family concerns to large firms. The three most important were

—General Telephone and Electronics (GTE)
(1983 turnover: 5.9 billion dollars in the field of telephony). The aggregated turnover of the GTE group was approximately 12.9 billion dollars in 1983, due to its diverse activities.

—United Telecommunications
(1983 turnover: 2.5 billion dollars)

—Continental Telecom
(1983 turnover: 2 billion dollars)

[N.b. As from 1991, Continental Telecom is now part of GTE. This makes GTE larger than any BOC.]

Although they had initially numbered around 7000, the field was substantially reduced as the sizes of individual companies grew due to mergers and regroupings which occurred from the 1920s onwards, later encouraged by the FCC.

By 1983, there were only 1459 independent companies, distributed as indicated in figure 4.

Missouri is a good example of the complexity of telephone penetration in a given region. In this territory operate:

State	Number of independent telephone companies in 1982	State	Number of independent telephone companies in 1982
IOWA	156	N. DAKOTA	22
WISCONSIN	108	LOUISIANA	22
MINNESOTA	91	ALASKA	22
TEXAS	75	MISSISSIPPI	21
ILLINOIS	58	VIRGINIA	20
INDIANA	52	KENTUCKY	18
NEBRASKA	49	MAINE	17
PENNSYLVANIA	48	MONTANA	16
MICHIGAN	48	FLORIDA	13
NEW YORK	45	IDAHO	12
OHIO	42	NEW HAMPSH.	11
KANSAS	42	WYOMING	10
MISSOURI	41	UTAH	10
OKLAHOMA	36	WEST VIRG.	9
GEORGIA	33	NEW MEXICO	9
S. DAKOTA	32	VERMONT	8
ARKANSAS	30	NEW JERSEY	5
ALABAMA	29	NEVADA	5
TENNESSEE	28	ARIZONA	5
N. CAROLINA	28	MASSACHUSETTS	3
OREGON	27	CONNECTICUT	2
S. CAROLINA	26	MARYLAND	1
COLORADO	25	HAWAII	1
WASHINGTON	24	RHODE ISLAND	0
CALIFORNIA	24	DELAWARE	0
		DISTRICT OF COLUMBIA	0
		Total	**1459**

Figure 4 Distribution of the 1459 independent companies before deregulation, by state.

—1 BOC: Southwestern Bell Telephone Company
—41 independent companies, of which the most important were GTE, United Telecommunications, and Continental Telephone.

The territories covered by each of them are tightly interwoven as figure 5 shows.

—*Competition in the long distance services, the "other common carriers" (OCCs)* From 1970—after the "MCI—Specialized Common Carrier" decision—the FCC authorized companies other than AT&T to set up their own interstate networks (in the case of MCI between Saint Louis and Chicago), in direct competition with AT&T Long Lines in the intercity market.

Known originally as Specialized Common Carriers (SCCs), these companies were commonly called OCCs, "Other Common Carriers". In 1983, they accounted for barely more than 8% of long distance communications market, valued at a total of 40 billion dollars.

The most important competitors to AT&T Long Lines were MCI (whose network served approximately 300 urban centres) and GTE-Sprint. Together they realised 80% of the market share of the OCCs. The numerous remaining companies shared the rest.

—*Competition in the other telecommunications services* In the other telecommunications services, different companies shared a market worth some 3 billion dollars, including:

—satellite transmission services, offered to "Common Carriers" by the Communications Satellite Corporation (COMSAT).
—domestic satellite communication services (DOMSATS) offered by such companies as Western Union, RCA, GTE-SPRINT, American Satellite, Hughes Communications, SBS, Vitalink;
—special link network operation services, offered by companies sometimes known as "Value-added Carriers", such as ITT, Domestic Transmission Systems, GTE, Tymnet, Uninet and Graphnet;
—cellular radio-telephony, where, in addition to AT&T and the "BOCs", Graph Scanning, GTE, Mobinet, and Western Union offered services.

● *Competition in the telecommunication equipment market*
In 1983, the American telecommunications equipment industry represented a market of the order of 40 billion dollars a year. The top American manufacturer of telecommunications equipment was the

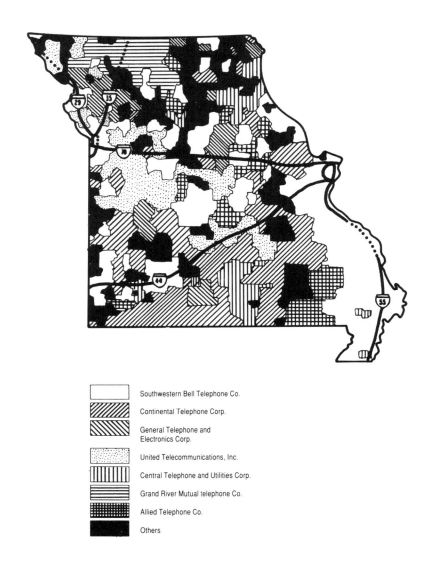

Southwestern Bell Telephone Co.

Continental Telephone Corp.

General Telephone and
Electronics Corp.

United Telecommunications, Inc.

Central Telephone and Utilities Corp.

Grand River Mutual telephone Co.

Allied Telephone Co.

Others

Penetration of telephone companies

Source: *Behind the Telephone Debates* C.L. Weinhauss and A.G. Oettinger

Figure 5 Missouri: company by company.
Penetration of telephone companies.

	Gross revenues (million $)	Subscribers (thousands)
AT&T	35 000	85 000
OCCs :		
MCI	1 520	1 550
GTE Sprint (1)	750	920
Allnet	180	150
ITT (2)	170	95
U.S. Tel.		80
SBS	70	90

Figure 6
(1) Born from the acquisition in 1983 of Southern Pacific Communications (SPC) by General Telephone & Electronics (GTE), another OCC which carried about 1% of the long distance traffic.
(2) The "world share out" agreement within AT&T occurred in 1925: after the government had required Western Electric to break away from its overseas subsidiaries, AT&T gave them up to ITT and NEC who, in return, undertook to not enter the telecommunications market in North America. This agreement lasted for 50 years

AT&T subsidiary Western Electric (annual revenue of 11.2 billion dollars, which was 23% of the world telecommunications equipment market). It accounted for 70% of the switching market (that is to say, exchanges, the heart of the system), 50% of the transmission market, and 25% of that of terminals. The company was also the top world manufacturer in this sector.

Certain foreign companies held a share of the American market which was not negligible, notably in the field of switching. First among these was the Canadian group Northern Telecom (second in the American market), then, of significantly less importance, the German company Siemens, the Swedish company Ericsson, the Japanese company Nippon Electric Company (NEC), and finally the English company Plessey.

It can be seen in 1983 that AT&T still exercized a quasi-monopoly in the field of telecommunications services and equipment. This was in spite of the attacks from the Department of Justice and companies attracted by new markets arising out of innovations in this sector (very often discovered by AT&T itself).

1.3 THE REAPPRAISAL OF THE STRUCTURE OF THE TELECOMMUNICATIONS MARKET

1.3.1 A "necessity"

It was the same telecommunications market which had appeared to be so stable and able to maintain its structure, which found itself called into question under the pressure of two factors: the desire for deregulation which had already manifested itself in other spheres, and the technological progress which had resulted from the convergence of computing and telecommunications. This is what prompted the FCC and the DOJ to act.

● *The influence of the economists of the "neo-liberal" school* represented chiefly by Milton Friedman and his "Chicago boys", started the deregulation movement in the United States. They advocated the elimination of government legislation which impeded the normal operation of "market forces", either because such legislation appeared unneccessary, or because it gave rise to counter-productive effects.

President Reagan made this philosophy his own on arriving at the White House. Four years later, he declared in his 1985 State of the Union speech that deregulation represented "the second American revolution; the revolution of liberty".

In this way telecommunications became the fourth sector to undergo the complete deregulation process, after banking and financial services; the distribution of petroleum products; and air, rail and road transport.

● *Technological progress* affected switching as well as transmission and terminals.

Telecommunications embodied more and more of the techniques of computing, and for its part, computing, relying on telecommunications to create inter-computer links, was being increasingly organized in networks.

—As far as switching was concerned, it was noticable that the convergence of telecommunications and computing increasingly underlined the inevitable evolution towards the Integrated Services Digital Network. This operating system would enable the simultaneous switching and transmission of voice, computer data, teletex and videotex.

—In the field of transmission, technological progress had widened the range of transmission media (coaxial cables, satellite communication, optical fibres) and at the same time increased the efficiency of each.

—As for terminal equipment, there too the integration of electronics and data processing technologies had led to an improvement of traditional equipment and to the development of new terminal equipment of greater and greater performance and diversification.

The burgeoning of these new technologies allowed the introduction of new services responding to an ever more pressing demand.

1.3.2   The redefinition of the market: new boundaries between services

The FCC, being aware of the issues of transformation in the telecommunications sector, and being responsible for its organization, brought out two successive enquiries in an attempt to draw a boundary line between the services which would remain regulated and those which would be deregulated, that is, open to competition. If "Computer Inquiry I" is of only historical significance, "Computer Inquiry II" was the start of the final process of deregulation.

● *Computer Inquiry I*
Launched in 1966, and completed in 1971, it was not put into practice until 1973. The main outcome of this inquiry was the creation of four distinct categories of service:

—traditional communications services;
—so-called hybrid communications services (telecommunications making use of computing techniques);
—computing services;
—hybrid computing services (computing services making use of telecommunications techniques).

In retrospect, it appears that the "hybrid" services did not have a sufficiently precise definition to be operational. The main result of this inquiry was that telephone companies were forbidden to provide computing services, including hybrid computing. If they wished to provide these services, they had to do so through separate subsidiaries (which AT&T, at that time chose not to do).

● *Computer Inquiry II: "deregulation"*
Technological advances, and above all the emergence of microcomputing, forced the FCC, after 1976, to launch a second inquiry: Computer Inquiry II.

This inquiry started in 1976 and was completed in 1980, but, due to practical difficulties, the outcome was not put into effect until January 1 1983.

Its main consequence was in the replacement of the Computer Inquiry I classification by a new one, dividing services into two categories:

—those known as basic services

—and those known as value-added (enhanced) services

Basic services were then defined as those in which speech or data transmission was brought about without any change to the form or contents of the information (for example circuit-switched data networks, special digital links, the switched telephone network).

Basic services remained regulated and all others were deregulated, that is, open to free competition.

Value-added services were all those which did not fall within the definition of "basic". For example, user-controlled message storage services, tele-consultation services (commercial exploitation of data bases, with videotex type services), information processing services (file creation, data processing, word processing). In a general sense, it concerned all those services aimed at offering the user specifically tailored services (message systems, transactions).

In the same way, Customer Premises Equipment, CPE, was deregulated. From now on such equipment was to be sold and no longer only rented, and to be marketed as free-standing items (rather than being tied to a particular service).

An overly brutal implementation of deregulation would, however, have upset the private terminal market. The FCC therefore decided that, as an initial measure, subscribers' terminals that had already been installed (as well as those that had already been fabricated) would remain regulated and "tariffed", and only those made after January 1 1984 would be sold as "deregulated" and "detariffed".

All of AT&T's terminals would have to be deregulated and "detariffed" by 1987 at the latest. On the other hand, the independent telephone companies had until 1991 to achieve the same.

In the same vein, in order to prevent AT&T and the 22 companies making use of its networks from abusing their power, Computer II required them to create independent subsidiaries (involving complete separation of staff, offices, computer resources, research centres etc.).

In addition, BOC customers were not obliged, as was necessary to reach OCC networks, to dial several additional digits in order to be connected to the AT&T long-distance network. This is why the FCC negociated with AT&T in order to arrange its tariffs to be slightly greater than those of its competitors until all the conditions for real competition had been set up. AT&T was also obliged, before 1986, to put into effect "equal access", that is to equip the whole of its network with the means of allowing the subscriber of any long distance carrier to be able to dial the same number of digits and to receive the same quality of connection as by using AT&T long distance services.

The FCC also imposed service tariffs which reflected the economic cost. Up until then, AT&T and all other telephone companies had systematically "subsidized" their local communications (whose tariffs were less than the real cost of the service) by the profits made on long distance communications. AT&T was therefore forced to impose an added price for BOC local communications, a significant price reduction (around 20%) in its long distance rates, and to levy access charges intended to cover the fixed costs associated with the purchase, by local companies, of the equipment needed to connect them to the interstate network.

The FCC had to review its position and, at the end of July 1983, it decided that, the access charge for individual users would be fixed in advance (2 dollars in 1984, 3 dollars in 1985, and 4 dollars in 1986). For business users, it would rise to 6 dollars during the first three years.

Because this charge did not cover all of the fixed costs that it was intended to cover, the remainder would be paid by the long distance carriers in proportion to the traffic carried.

This example is highly representative of the atmosphere in which deregulation was carried out, and of the way in which each party (States, telephone companies, users, Congress) attempted to solve their problems as they arose. This was without a pre-established plan, and in a manner that could be considered either as pragmatic or disorganized.

However, in order to create the conditions for real competition in the new "deregulated" telecommunications-computing environment, all parties worked together to set in motion the breaking up of the giant AT&T monopoly.

1.4   THE SHOCK OF THE BREAKING UP OF AT&T

In tandem with the deregulation of the telecommunications sector, and in order to create genuinely competitive conditions, the powerful AT&T was dismantled.

The break up was the result of an amicable agreement between AT&T and the Department of Justice (Consent Decree) in 1982. This decree represented a modification of the Consent Decree of 1956, hence the name MFJ (Modified Final Judgement). The MFJ was the result of an anti-trust action filed by the Department of Justice and dealt with by Judge Harold Greene.

On January 1 1984, AT&T was obligated to relinquish all interests in local traffic services, that is the 22 Bell Operating Companies. These BOCs were grouped into seven independent regional holding companies (RHCs) of similar financial strength (each having an annual revenue in the order of 10 billion dollars). Each one of them held a quasi-monopoly of local telecommunications services in its respective territory.

Before the break-up

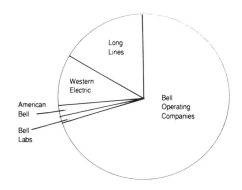

155 billion dollars

After the break-up

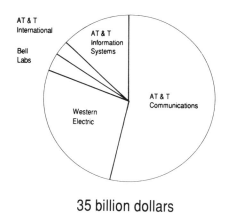

35 billion dollars

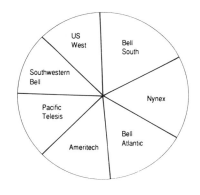

120 billion dollars

The sectors are proportional to the assets of each company

Figure 7

They were known either as RBHCs (Regional Bell Holding Companies) or, by association with the former terminology, RBOCs (Regional
Bell Operating Companies).

The Bell system counted at that time some 1 000 000 employees.
Following dismantlement, AT&T kept 375 000 of them, and the
others were divided among the RBOCs.

To counterbalance this, AT&T retained Western Electric and
became free to develop in fields outside telecommunications, notably
in the computing sector, and also to develop overseas markets.

After a complete restructuring of their allocation (figure 7), AT&T's
assets decreased from 155 billion dollars in 1982 to 35 billion dollars
in 1984.

The seven newly created RBOCs regrouped the former BOCs in
the following way (figure 8):

—NYNEX combined 2 BOCs and covered the north-east of the United
States;

Figure 8   Seven new independent and competing companies.

1 - PACIFIC TELESIS

Corporate headquarters: San Francisco
Gross revenues 1984: 8.5 billion dollars
Profits 1984: 970 million dollars

2 - US WEST

Corporate headquarters: Englewood (Colo.)
Gross revenues 1984: 7.8 billion dollars
Profits 1984: 950 million dollars

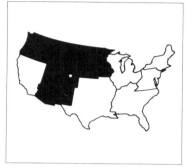

3 - SOUTHWESTERN BELL

Corporate headquarters: St Louis
Gross revenues 1984: 8 billion dollars
Profits 1984: 1 billion dollars

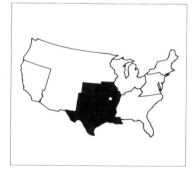

4 - AMERITECH

Corporate headquarters: Chicago
Gross revenues 1984: 9 billion dollars
Profits 1984: 1.1 billion dollars

Figure 9   Territories and financial strength attributed to the RBOCs at the start.

5 - BELLSOUTH

Corporate headquarters: Atlanta
Gross revenues 1984: 10.5 billion dollars
Profits 1984: 1.5 billion dollars

6 - BELL ATLANTIC

Corporate headquarters: Philadelphia
Gross revenues 1984: 9.1 billion dollars
Profits 1984: 1.1 billion dollars

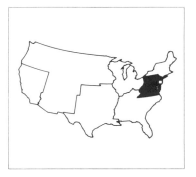

7 - NYNEX

Corporate headquarters: New York City
Gross revenues 1984: 10.4 billion dollars
Profits 1984: 1.1 billion dollars

Figure 9   (*cont.*)

—BELL ATLANTIC combined 7 BOCs and covered Washington DC and its neighbouring States;

—AMERITECH (American Information Technologies) combined 5 BOCs and served the centre-east of the United States;

—US WEST combined 3 BOCs and served 14 States in the centre-west of the United States;

—SOUTH WESTERN BELL which corresponds to the former BOC of the same name and covered the south-west of the United States;

—PACIFIC TELESIS GROUP combined 2 BOCs and covered California and Nevada;

—BELL SOUTH combined 2 BOCs and served the States of the south-east of the United States.

Each of the RBHCs covered a predetermined region of different size but with each region having a similar level of annual revenue and profit at the outset (figure 9).

Also, as part of the MFJ, AT&T sold its stock in the Bell system companies of Southern New England Telephone and Cincinatti and Suburban Telephone.

1.5   THE OPERATION OF THE AMERICAN MARKET FOLLOWING DEREGULATION AND THE BREAKING UP OF AT&T

At the time that AT&T was dismantled, the number of telephones in service in the United States was estimated at 250 million. Following deregulation, many of these were bought by their users. The phones operated over 110 million lines, which carried more than a billion calls per day.

1.5.1   Local services: a new definition

Deregulation called for a new definition of "local" services and "long distance" services.

Access, which continued as a "monopoly" for local services, was provided to the users by the local operating companies (22 BOCs)— now grouped into seven RBHCs, and was "regulated" by the states. The FCC was concerned with tariffs and access to interstate services and "deregulated" or competitive services as defined in "Computer Inquiry II".

Local services were also provided by the 1 400 independent companies to whom specific areas for monopoly operation had been franchized along the way as the telecommunications story in the United

States unfolded. Their size varied widely, ranging from GTE, with more than 10 million lines, to Island Telephone Company which only had 28 lines. On the whole, these companies functioned like the BOCs, but with a somewhat more lenient degree of regulation.

The number of independents continuously diminished, with economic pressures forcing regrouping. A dramatic example of consolidation is the GTE acquisition of Contel in 1991. This regrouping resulted in GTE becoming the largest telephone company (superseding the largest RBOCs). Excluding the two largest companies (GTE and United Telephone Systems), these independent telephone companies repesented only 10% of the total number of lines.

In summary, historically each local BOC or independent operating company was franchized by the states a geographically defined zone in which it was able to provide, in a monopolistic regime, a certain number of services. In this way, the subscriber did not choose his telephone company, but it was imposed upon him as a function of his geographic locality.

However, these franchized area services did not really represent the "post deregulation" definition of so called local or basic service.

The AT&T Company, having lost its local service, exacted a new definition of local service through the MFJ. As a result, the territory of each BOC was itself sub-divided into Local Access and Transport Areas (LATAs).

The American territory was thus subdivided into 161 LATAs.

The BOCs and the independents, that is the local operating companies, were only allowed to carry those communications located within a single LATA (Intra-LATA), even if they had several LATAs in their franchized area of local service monopoly; communications between LATAs (Inter-LATAs) were provided by the long distance operators such as AT&T Communications, GTE Sprint and MCI.

The users, as in the past, did not have the option of choosing their local service company, but now could choose their long distance carrier for InterLATA and Interstate long distance service.

As a result of one particularly important measure imposed by the Greene Court and the Federal Government in the MFJ, the RBOCs pursued their research effort through the intermediary of a specialised research company, "Bellcore", which belonged to all of them in common. The level of modernization in the seven zones was not equal.

1.5.2 Bellcore

A product of the breaking up of AT&T, Bellcore (Bell Communications Research) was born on January 1 1984. Its creation was a condition stressed by the Judge Harold Greene: "Bellcore is there to verify that the whole American network functions as one single and unique network having the highest possible quality". (US District Court June 30, 1987).

Not exclusively at the service of the seven RBOCs, Bellcore had a double mission:

—to be a research company at the service of the seven RBOCs in a competitive environment: with contracts placed either by all the RBOCs in common, or by just one or a group of RBOCs.
—to serve as a coordination centre for communications services with regard to national security interests and emergency services activities. It was also to contibute to the standardization of methods, materials and hardware.

Bellcore treats the RBOCs as customer companies: RBCs (Regional Bell Clients). It offers two types of facilities:

—those concerning the "infrastructure" ("core projects") in which all the RBOCs are involved and which represent in 1988 some 37% of the budget.
—At the same time those called "specific" ("elective programs"), which represent 63% of the budget, and which could be either multi-customer projects or projects concerning a single specific customer.

Bellcore's 1988 budget was of the order of one billion dollars, and its workforce was 8 200 strong.

The RBOCs represented more than 90% of Bellcore's resources in 1988, a situation from which the company is trying to diversify: two non-affiliated companies (CBI, SNET) 3%, Licences (software, teaching) 2%, Others (Bellcore suppliers) 4%.

This arrangement offered the following advantages:

—the economy and synergy which accrue to the RBOCs from having a shared research resource,
—the quality of Bellcore "knowhow", initially coming from the famous Bell Labs.
—standardization of suppliers goods and materials, the search for solutions to the national need for normalization.
—the search for solutions in questions of integration and system architecture.

AT&T, which had retained its own research capacity with its slimmed-down Bell Labs., remained the uncontested leader in long distance communication, and the manufacture of equipment.

1.5.3 Long distance services

This market was covered by the new AT&T and by all other companies that wanted to participate, since, in principle, the sector was totally open to competition.

It is, however, important to note that the new AT&T, having a distinct advantage at the outset, continued to be regulated as the "dominant carrier" by the FCC in matters concerning tariffs and services, until it could be confirmed that real competition had been established in the market.

The AT&T long distance network represents 1.4 billion km of links based on cables, radio, satellites and optical fibers, and is connected by 167 central nodes.

● *The new AT&T*

The new AT&T (since January 1 1984) has been composed of two main sections:

AT&T CIS (Communications & Information Systems), and AT&T Technologies (figure 10).

AT&T Communications and Information Services supplies a wide range of services regulated by the government including:

—basic long distance services,
—telecommunications on leased lines,
—"800" free call services,
—"900" special services
—a high speed digital network intended for businesses.

AT&T Technologies supplies products and services without having the restrictions of the pre-1984 regulations; it develops, manufactures and markets a large range of telecommunication and computing products. The company employs 200 000 people. This industrial strength was substantially augmented by the acquisition of NCR, a major computer manufacturer, in 1991.

● *More than 500 companies offer long distance services in competition with AT&T*

These companies include the "Other Common Carriers" (OCCs): companies which rent capacity on the AT&T long distance networks for part of the route, and complete it with their own satellite, radio or optical fiber links; the WATS resellers: companies buy transmission capacity in bulk from network operators in order to re-offer it to customers at lower prices. Sometimes the WATS resellers also offer value-added services.

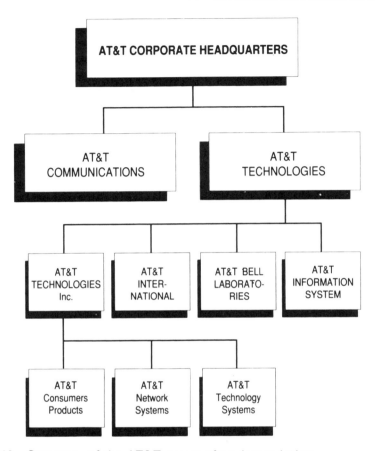

Figure 10   Structure of the AT&T group after deregulation.

The list of services offered by these companies is very large: long distance services using terrestrial links, satellites, and/or radio "bypass" routes; integrated private voice circuits, data, telefax and videoconferencing, personal radio-paging, mobile cellular systems, electronic mail, videotex and data base access and software and voice messaging.

The two real competitors to AT&T (which, with a 35 billion dollar annual revenue in long distance services held 63% of the American long distance market and was at least 10 times larger than any of its competitors) were MCI and US Sprint (a subsidiary of United Telecom); they respectively held 17% and 12% of the American long distance market at the end of 1989.

	1985	1986	1989
AT&T	67.8%	65.3%	63%
MCI	5.3%	6.6%	17%
US Sprint	2.9%	3.8%	12%
Autres	24%	14.7%	8%
	100%	100%	100%

Figure 11   Evolution of the long distance market share.

A subtle game of lobbying to obtain markets was played out between MCI and US Sprint, and not without success! For example, in July 1989, US Sprint landed 1/3 of the huge Federal government market, with the other 2/3 accorded to AT&T.

In the long term, the small independent carriers and resellers will not all survive, or in any case not in the same form. Mergers and acquisitions will gradually reduce their number.

For the time being, dozens of companies are managing to survive in the long distance market, but it is clear that only a few will eventually remain: the three largest AT&T, MCI, US Sprint and probably ALLNET and Williams Communication, plus around 200 WATS resellers.

● *Private networks, LANs (Local Area Networks) and long distance*
A large number of businesses wish to avail themselves of an integrated services network which allows them to transmit voice, data and images, and to be capable of linking their local installations within the American territory and beyond.

AT&T, the RBOCs, and the OCCs are not capable of meeting this requirement on their own. According to the National Telecommunications and Informations Administration (NTIA), 43% of the largest 500 American companies, which account for 17% of intercity traffic, already make use of their own network, for example General Motors, with its specialist subsidiary EDS (Electronic Data Systems), the banking networks like that of CitiCorp., the aeronautic firms such as Boeing, or even the Sabre network for airline reservations.

Businesses are not hesitating to acquire such networks to meet their needs.

This market is far from negligible. It already represents 10% of the telecommunications market, and is continuing to expand.

These private networks represent income, not only for the companies who install them, but for the long distance carriers and local companies who receive a fixed income by leasing out specialized lines.

Today, more than 100 installation companies can be counted in the United States, which represent approximately 70% of the world market in local networks, with some 100 000 installations. Cable TV companies and satellite companies such as Teleport have provided fiber optic facilities directly to large users and bypass the local company entirely.

The LAN market is divided into segments. The most active sector, naturally, is that of commercial offices (60% of the market), but other sectors are developing, like that of computer-aided design (10% of the market) and factory automation.

There is also a demand for private short distance point-to-point links which bypass the local service companies. This demand is increasing by 40% a year.

In addition, miniature earth stations (very small aperture terminal "VSAT") for professional use are also expanding at 45% per year.

Some of the newcomers are the private long distance network operators, such as Telecom Network. This company (a consortium of regional fibre optic firms: Consolidated National, LDX National, Litel, Microtel, Southern Net, Southern Fibrenet, Wiltel) has operated a national network since 1985.

Alternative private circuits to link different facilities of a single company are also available by satellite (VSAT). As an example, WALMART, a distribution company for business centres, has interconnected 1114 points of sale, exploiting the largest private satellite network in the world.

The arrival of IBM and AT&T on the market has been decisive as far as the definition of standards is concerned. Indeed, with IBM introducing telecommunications into its computer products, and AT&T computerizing its telecommunications products, the stance of the smaller providers has been to fall into step with one or the other, and sometimes both.

In this new and continually evolving market, the number of manufacturers continues to increase. Accordingly we find Apple, Apollo, AT&T, Bridge Communications, Codex, Datapoint Corporation, Hewlett Packard, Intel, Motorola, Bradley Information Systems, Network Research Corp., Symtek, Ungermann-Bass, TRN, Wang, Xerox, 3 COM Corporation and Bull HN.

Compounding this diversity of means of communications is the explosion of wireless communications. This was initiated by the introduction of cellular radiotelephone at about the same time as the divestiture of AT&T. The FCC defined market areas throughout the nation and mandated two licensed operators in each market area; one wireline or telephone company, and one non-wireline operator (not a telephone company). The number of competitors for this business far exceeded the number of licences available. With FCC's urging, the process was expedited by the joining of several proponents into a single consortium application. The small and rural areas were awarded by an FCC lottery. This whole process, which took from 1984 until the end of the decade, has resulted in a large and still growing unregulated communications industry.

The success of cellular and the technological advances of miniaturization and digital encoding have brought forth a number of new services for cordless and personal telephones—one such development is the laptop computer with built-in radio capability for sending Fax and data.

1.6 A PERMANENT REGULATION

The regulatory process is constantly evolving. American legislation is striving for a market which is increasingly competitive and responsive to the needs of the users.

The questions raised concern access, the quality of the network, the availability of technologies, and the equality of treatment of providers and users.

A constant restructuring of the market has resulted, at both the national and the local level.

● *1985: Computer Inquiry III or a "new American proposal for the opening up to competition"*
Structural separation, that is, the constraint of only being able to offer value added services via structurally and completely separate subsidiaries which was imposed on the Bell Operating Companies by Computer Inquiry II, turned out to be restrictive and costly for them. They agreed that, far from facilitating competition, these measures were unecomonical and contrary to the interests of consumers at large. Under pressure from these companies, the FCC therefore proposed a new inquiry; Computer Inquiry III. Set up in 1985, it attempted to improve on Computer Inquiry II by establishing "new deregulation rules" and, notably, by abandoning the arrangements concerning structural separations. Four other, so called "non-structural", guarantees were substituted.

1) *The most restrictive, the Open Network Architecture (ONA):* a technical concept of a marketing nature.

In order to supply Value Added Services, the BOCs had to each put into effect an open network architecture (ONA) plan which was subject to approval by the FCC.

There were two aims of this new architecture:

—To "break down" the network into "elementary components" so that any user of "Value Added Services" (supplier or final user) would be able to connect to the basic network at any point;
—To allow non-discriminatory access by all users of these services.

The BOCs were also required to offer to others the "same services" as they used themselves, with the same quality, the same reliability, the same cost... :CEI (Comparably Efficient Interconnexion).

The ONA therefore strives to allow the network to evolve so that new value added services can be created by providers other than the telephone companies without the need to conform to any particular restrictions or to incur unwarranted costs.

This had the objectives of preventing the BOCs from abusing a position of dominance which could give them privileged access to the network, and allowing the dynamic development of "Value Added Services" by all providers on an equal basis.

Naturally, the ensemble of networks involved in these projects would need to be easily inter-connectable amongst themselves. The FCC requested that each RBOC formulate a new plan taking the best features from those of the seven others. The eight plans were thus quite similar. In practice, a single American network was thus created, even if its components differed as a function of the particular needs of each category of BOC customers.

The ONA plans were ordered to be submitted by February 1 1988.

This was accomplished. However, in December 1988, the FCC declared them to be less than satisfactory and ordered new plans to be submitted on May 19 1989. These were approved in April 1990 subject to certain amendments.

2) *The BOCs* had to publish their system of cost accounting.

3) *The BOCs* had to meet requirements of customers regarding access to their own information.

4) *Other requirements* concerned the publication of information about the network.

These FCC rulings and the pressure of the industry coupled with strong interest on the part of several State regulators has moved the degree of "openness" to co-location of other providers" equipment within BOC central offices. This co-location was seen as the only way to achieve the objectives of ONA and minimize the issues of equality of costs.

● *April 1989: the public pay-phone market was also opened up to competition*
From April 1 1989, the market for public telephones, was opened to competition.

Following the Judge Greene ruling of October 1989, owners of premises containing public telephones were given approximately six months to designate the long distance operating company which would serve them.

Two thirds of them made their selection. 65% chose to remain loyal to AT&T; 10% chose MCI, 8% US Sprint, and 7% chose other companies, notably Alternative Operator Services Companies (AOS)(1). Those who did not specify an operating company were arbitrarily allotted one. The process left unchanged the proportion of market share for each carrier arising from the stated preferences.

Among the questions which arose was the technical method to be adopted in order to allow public telephone users the freedom of choice of access to any long distance company: the simple use of a prefix (which would render a good number of telephones obsolete), or recourse to a free number giving access to the chosen operator (prolonging the call connection delay noticeably)?

Another was, how to avoid billing when the party being called did not respond (several operators based their mechanism for instigating charges on the elapsed time from the beginning of the call routeing process).

And another: How to ensure that the billing would be calculated as a function of the distance separating the caller and the callee, and not the operator (who could be situated at a third place at the other end of the country)?

Several such questions remain without immediate solution and stand as a good example of the complexity of American telecommunications and the issues which confront the FCC and the Greene Court. At the last juncture, public telephone and operator services providers have been ordered to permit access to the carrier of the

---

(1) Initially, the AOS offer operator assistance services, and these have had higher prices in general than the main long distance network operating companies. They justify this difference by the number of additional services that are available.

International Telecharge is a case in point: its operators can reply in 24 languages, calls can be billed on credit cards, and a very popular messaging service is offered: if the line is engaged, the caller is invited to record a message of one minute duration; the required number is then regularly re-dialled, and when it is finally free, the pre-recorded message is automatically delivered to the called party. Recently (1991), the prices of the AOS have trended down to the general level for the individual services.

user's choice via a free standard dialing format. What remains unresolved is how the public telephone providers will recover their costs for public access to the network.

Operator services providers have been ordered to announce the name of their company and to make available their prices to the public.

● *Relaxed regulatory restrictions for AT&T:*

—July 1989: the FCC adopts the "price cap" for AT&T   Since 1989, the FCC has permitted AT&T to fix its tariffs with increased flexibility by freeing it from the "rate of return" restrictions.

Until this date, the regulation of carrier's tariffs had been controlled by limiting the of rate of return of the total business of the carrier, fixed at 12.75%. This principle encouraged cross-subsidy, thus giving AT&T a clear advantage for the launching of new services, compared to its less powerful competitors. In addition, the rules discouraged the carrier from improving the productivity of his management, since any gain in profitability had to be directly passed on to the users.

From 1989 on, the rate of return of the firm was no longer limiting. Price control was from then on effected by means of the "price cap".

The annual increase in the capped tariffs should simply be at least three percentage points less than the inflation rate. This 3% adjustment corresponds to 2.5% annual productivity gains achieved by AT&T, and the remainder is a "gift" to the users.

In addition, to prevent AT&T from distorting the basis of competition by underpricing certain services, a lower cut-off has been fixed for price changes (5% below the "price cap"; that is 8% below the rate of inflation). Any tariff set less than this cut-off is submitted for special authorization by the FCC.

The prices of capped services are thus entirely free to vary between 3 and 8 points below the official rate of inflation.

However, in order to retain some flexibility, these rules apply to three "baskets" of services, taken together: general public, businesses, number 800 free-call services. Each basket must stay within the constraints described above, but each service, individually, is limited to five points above the "price cap" (four points for off-peak residential tariffs).

On the other hand, AT&T will now be able to change its prices much more quickly: each new tariff can now be submitted to the FCC 14 days in advance, not 45 days as was previously the case. The 45 day period had allowed competitors to market new offers even before those announced by AT&T.

Above all, this decision has a key symbolic importance. The FCC, in relaxing the regulatory restraints imposed for a period of five years, implicitly recognized that AT&T is no longer an unassailable giant. Its competitors have gained sufficient strength, and it can from now on be left these freedoms in the competitive environment, without undue risk of it reforming a *de facto* monopoly as a result.

The abolition of "rate of return" regulation for AT&T, resulting from a two-year battle between the chairman of the FCC and certain members of Congress, constitutes an additional step in the process of deregulation.

—September 1989, AT&T is authorized to enter the electronic publishing market　By a decision on August 24, Judge Greene lifted the five year MFJ restriction which, since the breaking up of the Bell system in 1984, barred AT&T from electronic publishing. AT&T will now be able to offer, across its network, a large range of information services such as stock market data, videotex and news bulletins. It will be able to offer services in direct competition with the "electronic publishers" such as Dun and Bradstreet, Dow Jones, McGraw-Hill or Reuters.

The restriction on electronic publishing was originally justified as necessary to protect the information technology industry against unfair competition from AT&T, in view of its domination of long distance communications. The judge justified his decision on the grounds that real competition now existed in the long distance communications market; a monopoly, or even a quasi-monopoly, no longer existed.

The entry of AT&T into "electronic publishing" was favorably received by professionals in the information services area who expected considerable development in this market, as well as interesting opportunities of partnership.

However, the Regional Bell Operating Companies (RBOCs), who were aiming at the electronic yellow pages market, did not benefit from the same flexibility. They were still prevented from offering electronic publishing services, Judge Greene arguing that they would constitute a virtual monopoly.

—1991, Greene denial of information services to RBOCs undone by Appeals Court　The US Circuit Court of Appeals, upon the review sought by the RBOCs, sent the question of releasing the RBOCs from the MFJ restrictions against information services back to the Greene Court. Judge Greene then agreed to lift the ban. His July 1991 order was accompanied by his opinion that it was the wrong time to lift the ban, but that the higher Court had made any other decision indefensible.

He stated that, although he believed the result could be anti-competitive, he could not prove that, under the standard imposed on him by the April 1990 Appeals Court remand.

At the instant of Greene's order to lift the ban, he also imposed a stay of his order to give those against the release the time to pursue appeals. This lengthy process is the state of the matter at the time of writing (October 1991).

1.7   PROGRESS OF AMERICAN REGULATION AFTER THE BREAK UP OF AT&T

The structure of American telecommunications continues to attract the closest attention of Congress and the Administration.

● *June 1990: The reconsideration of Computer Inquiry III by the Federal Court of Appeals*
During 1990, numerous legislative and regulatory questions were raised.

The FCC affirmed its precedence over the States in its "Computer Inquiry III" findings. Regarding matters in the field of value added services, it treated these services as peripheral to the main telecommunication services. In addition, it removed the requirement that the RBOCs create structurally separate subsidiaries as a condition for their providing value added services. It substituted instead accounting separation through the use of a cost manual.

For this move, it was immediately and strongly accused of exceeding its authority by the State regulatory authority, and by the Public Utility Commissions (PUCs), who were unhappy at being disposessed of an important part of their regulatory powers. It was also contested by the companies competing with the RBOCs in the value added services sector.

The more dynamic among the concerned parties challenged the legality of the Computer III decision of the FCC in the U.S. Court of Appeals. This involved the State of California, the PUC of California, MCI Communications Corporation and Pacific Telesis.

The Court of Appeals gave its verdict in June 1990: Computer Inquiry III was for the time being suspended. The Court had in effect held that the FCC had not sufficiently justified its Computer Inquiry III rulings.

In the first place, the Court ruled that the FCC had not demonstrated that market conditions had sufficiently changed to allow the the RBOCs to dispense with the formation of separated subsidiaries before offering value added services. Furthermore, according to the judgement of the Court, ONA did not preclude cross subsidy.

Finally, the Court of Appeals did not recognize the FCC's right to usurp the regulatory prerogatives of States which was implicit in its opposition to the establishment of stricter rules, and in its intervention in the regulation of intrastate value added services.

The court left the FCC the opportunity, however, of presenting a better argued case in one year's time.

In the meantime certain RBOCs had already started to offer value added services under the "Comparably Efficient Interconnection" ruling without separated subsidiaries, and accounting separation of costs. What should they do? Should they re-organize themselves once again? Should they wait?

The FCC was concerned at this backward step at a time when the European Community was itself starting along the road to deregulation. To quote its chairman Sikes, in July 1990, "it is ironic to realize that the Court is preventing us from forming a national regulatory framework at the very time when the EEC and Japan are moving in the opposite direction...".

The precise consequences are difficult to determine. It will be necessary to wait for a new formula ... Computer IIIbis!

● *Should Congress relieve the BOCs of the MFJ restrictions?*
On April 18 1989, the seven Regional Bell Operating Companies (RBOCs) jointly submitted to Congress a proposal for legislation authorizing them to manufacture equipment and provide information services.

The provision of long distance services, equipment manufacture, and information services are the three major telecommunications activities from which the RBOCs are presently prohibited by the MFJ. These restrictions had been included in the MFJ to enforce the anti-trust laws by preventing the break up of AT&T from leading to new overly powerful telephone monopolies.

For several years, the RBOCs have been conducting a major lobbying campaign in order to bring about a review of the MFJ. These restrictions of the MFJ have also been criticized by the Department of Commerce for their negative effects on the industrial competitiveness of American telecommunications enterprises.

The RBOCs proposed two major amendments to the 1934 Communication Act, to:

—give the RBOCs the right to supply information services, on condition that the market in any one State would not be monopolized;

—permit them to develop software and manufacture telecommunications equipment, on condition that their procurement procedures should not be discriminatory.

However, the legislative proposals of the RBOCs did not address the question of long distance telecommunications services, nor that of cable television networks, which are also MFJ prohibitions.

At the end of 1989, powerful members of Congress agreed to "modify" rather than "lift" the restrictions affecting the BOCs defined in the MFJ. These proposals could open the way to information services for the BOCs.

The FCC expressed the view that relaxing the constraints on the BOCs would eventually benefit the public, and that regulatory policy should not systematically oppose it.

With respect to the manufacture of equipment, the RBHCs sought the right to establish a foothold in this sector, but a concensus of opinion emerged against them. The BOCs sought relief from the broad definition of Judge Greene who had prohibited the development of products in the United States, while not interfering with the manufacture of equipment or the development of software for telecommunications systems abroad, as long as these innovations would not be reintroduced on the American market. This restriction appeared to be uneconomic in the eyes of the RBOCs and their supporters. After examining the risks of cross-subsidy, vertical integration, discrimination, reserved markets and of uncompetitive practices, the legislators hoped to improve this situation.

Throughout this debate, equipment manufacturers opposed the RBOCs, pointing out that the MFJ had been adopted in the spirit of the anti-trust laws against the vertical integration of the Bell-System, and that nobody wanted a return to the former situation. But was their lobby powerful enough?

By mid-1991, the Senate passed a bill to remove the ban on manufacturing, but included domestic production (content) requirements. Also, the House subcommittee is considering two bills removing the manufacturing ban.

While the debate in Congress struggled with a balance between opposing interests and no action, events in the courts suddenly brought about change. Judge Greene's denial of the RBOCs request to enter the information services sector was remanded to court by the Court of Appeals.

The Appeals Court found that Judge Greene's basis of denial was not based upon the correct legal standard. As a result, Judge Greene issued an order in July 1991 allowing entry into information services. Further, he ordered an injuction to allow interested parties a further appeal. However, the Court of Appeals promptly removed this injunction.

In this manner the courts removed one of the major MFJ restrictions without imposing any new terms or conditions,

● *Should long distance telecommunications be completely deregulated?*
AT&T denies being a dominant operator at the present time, and has demanded treatment equitable to other carriers from the FCC.

In the view of the FCC and its chairman Sikes, the restrictions imposed on AT&T should be effectively limited, because there is sufficient evidence that the long distance market is now competitive.

● *Should competition be introduced into local communications?*
The FCC believes that the BOCs have not invested enough nor modernised the networks sufficiently in the past few years. In order to prod them into increasing their level of investment, the FCC has removed rate of return regulation and subjected them, like AT&T, to a "price cap" for interstate access services. A separate issue is the possibility of introducing competition in local services, for example by means of cable network operators. In some states, this is now underway....

1.8 TODAY'S SERVICES AND EQUIPMENT MARKET IN THE UNITED STATES. PERSPECTIVES ON THE 1993 HORIZON

The essential consequence of deregulation has been the opening of the equipment and services markets to competition, with the goals of adapting to the new needs of specific customers, of evolving solutions which respond to the demand for easier residential use, of decreasing prices, and above all of increasing productivity for business customers.

It has resulted in a transformation of a closeted market into one of multiple buyers (no longer limited to AT&T, but including all the BOCs) and multiple vendors (no longer only, or nearly only, Western Electric, but now including all firms who sell equipment of interest).

An upheaval of the industry has been brought about; subsidiaries formed, joint ventures initiated, and strategic alliances constituted.

1.8.1   Services

● *Local telecommunications*
Local telephone service, which is practically monopolistic, is still by far the most important among the telecommunication services. In

fact, it earned almost 100 billion dollars in revenue at the end of the decade, in spite of the ability of large enterprises to bypass local networks and divert their traffic, which amounted to approximately 3.8 billion dollars of revenues lost to the RBOCs.

The RBOCs on their own, however, earned 80 billion dollars in revenues at the end of the decade.

They therefore have considerable funds at their disposal, which is why they have elected to diversify in both the United States and abroad.

Which areas are these companies putting their effort into? The following notes give some particular examples.

—*Personal radio-paging, radio-telephones*   BellSouth, already a leader in this domain in the United States (1 million users), is also active in Australia; Advanced Communication in France, participating in Datech with the operator billing system of the TDF company, in Buenos Aires with a local consortium....

NYNEX, Pacific Telesis, Southwestern Bell, and US West also want to develop in this sector; not only in the United States, but also in Western Europe (particularly in the UK, Italy and West Germany) and in Eastern Europe, China, Thailand and Taiwan....

—*The publication of directories and associated services (publicity and printing enterprises)...*   Southwestern Bell, a leader in the United States, has also set up joint ventures in Israel (AUREC), Korea and Australia.

Ameritech has contacts in Thailand and Japan for English language directories, and is attempting to capture some of AT&T's markets in Asia. Nynex, which is already developing its yellow pages publishing activity in New Jersey, has decided to expand across the whole world.

—*Computing and telecommunications engineering*   Bell Atlantic has a very strong position in computer maintenance. Notably, in 1988, it bought BCE Canada's subsidiaries to form Sorbus International, which has wide penetration in Europe.

BellSouth is putting up "Smart Buildings" and local networks in Taiwan and Shanghai, and is installing digital networks in Guatemala. It has also reached an agreement with SIP (Italy) to start work on software for the Italian network.

Nynex has won a contract with British Telecom for a network integration system.

—*The sale of equipment*   Bell Atlantic, for example, distributes DEC equipment.

BellSouth has bought Data Service (Great Britain) which distributes IBM equipment.

Southwestern Bell is looking for development opportunities in this sector, especially with its British subsidiary (Freedom Phone), and has a distribution agreement with Bell Canada for telephones.

—*Cable television—videotex*   Pacific Telesis is very aggressive in the cable television market; the company even wants to launch into program production. It is currently building the Palo-Alto network.

Southwestern Bell is interested in high-definition television.

US West has 10% of Lyonnaise Communication, and 25% of Cable Corporation (Great Britain).

—*Other fields of diversification*   Bellsouth runs telemarketing operations and is participating in the teleport project in Metz (France).

Bell Atlantic has agreements with IBM and Siemens for voice and data networks.

Nynex is involved in property operations linking sale, information services and leasing. It has information and personnel exchange agreements with numerous PTTs (France, Netherlands, Italy, Spain, Scandinavia, Singapore and Japan).

US West has a stong presence in property, financial services, and compact discs.

—*Concluding notes*   Ameritech has up to now followed a policy of consolidating its business on its own territory, and seems to consider foreign markets as limited and unstable.

Bell Atlantic has opted for international diversification following two directions: computer servicing and network planning and organization.

Bell South has indicated a strong interest in international markets, and has focused on mobile communications.

South Western Bell shows great caution towards foreign markets.

US West is looking towards Poland and Hungary with cellular telephony, and towards the USSR with optical fiber.

Each of the RBOCs generates a level of cash flow which is among the highest in the United States.

● *Long distance telecommunications*

Long distance communications service is the arena of a growing competition between AT&T, MCI and US Sprint. The revenues involved are of the order of 50 billion dollars.

Because of this competition, there is pressure on prices, and this is why, although the trade volume is increasing, growth in income is only 3%.

—*Mobile radio communications*   Revenues from radio-telephone services rose to approximately 5 billion dollars by the end of the decade and should be of the order of 15 billion in 5 years time. The RBOCs, as we have seen, are among the main players in the market.

—*Satellite services*   This is a very competitive sector, where prices are currently negotiable, since there is excess capacity of approximately 30%.

Even more ruthless competition between the nine American satellite operators is expected in the future.

The global revenue in this market sector is of the order of 650 million dollars per year.

Satellite operators are concentrating their activities on private networks, direct broadcasting, backup communications and mobile services.

—*Value added networks (VAN)*   Telenet (a subsidiary of US Sprint) runs the main public data network with 41% of the market, followed by Tymnet (McDonnel Douglas) with 36% of the market, GEISCO (General Electric), EDS (General Motors) and Infonet.

The future will be marked by restructuring and the search for partnerships among the VAN service providers.

It is interesting to note that, in the United States, Videotex is not a mass information service. Its main applications are in banking, finance, marketing and insurance.

In the next five years, it is certain that global growth in revenues will be based on the offering of new services, on an increase in the volume of telephone traffic inside the USA, and also on international diversification.

1.8.2   Equipment

● *Network equipment*

The RBOCs, independent telephone companies, and long distance companies continue to modernize their networks.

The American market is dominated by AT&T Technologies and the Canadian Northern Telecom. The RBOCs are attempting to expand their supply base however and to bring down prices through competition. For this reason, NEC, Ericsson, Alcatel and Siemens in particular regularly increase their market share, although it is still small.

● *Optical fibers*
The American market in this field is currently worth about 2 billion dollars per year.

This is an enormous market which will develop further due to the RBOC plans for installation of optical fiber to the home. AT&T and Corning Glass dominate production. Among the foreign industrial powers in this sector, Siemens is in a good place.

● *Satellite systems*
The market for satellite transmission systems (satellites and earth stations) is relatively stable, on the order of 2.6 billion dollars per year.

A resumption of growth is expected, due to new orders by Intelsat and other operators.

There is European competition here. Alcatel Spain signed a contract worth 86 million dollars at the end of 1988 for the supply of equipment for 5 Intelsat VII satellites made by Ford Space.

● *Cellular radio-communication equipment*
At the end of the decade, the market was worth approximately 2.5 billion dollars, and there were 2.5 million subscribers.

The industry remains concentrated in the hands of a few manufacturers: Motorola, AT&T, Ericsson, and Northern Telecom. A fifth manufacturer, Astronet, is now establishing itself in the market.

With regard to terminals, imports have a value of approximately 400 million dollars, mainly coming from Japan, South Korea, Hong-Kong, Singapore and Taiwan. The annual growth rate is more than 50%.

Forecasts predict an increase to between 12 and 20 million subscribers by 1995.

● *Private terminal equipment*
This market has a value on the order of 5.8 billion dollars per year including PABXs, intercom systems, telephone and telex terminals, telefax machines, modems, and local network equipment.

Local network equipment sales are experiencing strong growth, (20%), with production approaching one billion dollars (against only 100 million dollars in 1982).

PABX production increased by 3% in 1990. On the other hand, the production of telephones and intercom systems dropped by 3%

in a market worth approximately one billion dollars. AT&T, which holds 35% of this market, is sustaining a continuous erosion of its share by Japanese and South Korean manufacturers. Also, the telephone companies continue to improve their alternate PBX-like service (CENTREX). This has been accelerated by the deployment of digital central offices.

1.9   OVERALL ASSESSMENT

The different steps towards deregulation undertaken since 1984 have largely convinced Americans that their market is now truly open and competitive, and that both European and non-European players are already present without any hindrance to their development. They therefore consider that, if any supplier offers better products at better prices, it will not be discriminated against. Accordingly, to the Americans, the decision of whether or not to enter the American market rests solely on operational questions and does not require political intervention.

This picture is at odds with the real situation. Industrialists or government agencies are reluctant to favour foreign companies who are making strong efforts to penetrate the United States, such as Ericsson, Siemens, Northern Telecom, Alcatel and NEC.

They consider that foreign companies supported, by their governments, should not be allowed to compete with American companies in the USA; all the more so since, rightly or wrongly, American firms unanimously feel that their access to the European market is difficult.

The conditions necessary for foreign companies to succeed in the American market are difficult to fulfil however. They must offer products conforming to American standards, be able to set up good distribution networks, and, if possible, to manufacture in the USA.

The barriers restricting entry into the American market are both real and psychological.

First come standards and the application of FCC rules. Obtaining FCC approval seems particularly complex for European companies.

In addition, a number of "unwritten" rules exist which lead to real, but invisible obstacles to penetrating the American market.

A foreign company, according to the "Buy American Act", must show that 50% of the contents of its product are of American origin. This implies long and costly procedures.

In addition, the law limits foreign participation in the operation of American networks to 20%.

An unofficial protectionism against entry into this market obliges companies to offer very innovative products, requiring a considerable investment.

The company which has profited the most from deregulation is undoubtedly the Canadian company Northern Telecom (a 53% subsidiary of Bell Canada). Since 1971, it has had manufacturing activities in the United States through its subsidiary Northern Telecom International. The RBOCs welcomed this company after divestiture as an alternative to AT&T, but have not moved significantly beyond the diversity.

The cost of establishing a presence in the American market is high and needs to be based on a long-term strategy; Ericsson has stated that it devoted 9 years and spent 20 million dollars in order to adapt its PABX MD 110 for the American market; Northern has indicated that it invested more than 30 million dollars in setting up its network for businesses.

Deregulation in another way has convinced the American industrialists to invest in foreign markets. They want to compete not only in the American market, but also on a global basis.

In 1988, the Congress passed a "Trade and Competitiveness Act". This Act invested the President with important bargaining authority to open foreign telecommunication markets. The "Trade Act" stipulates that the disregarding of a trade agreement (for example a commitment by a foreign country to open its internal telecommunications market) would lead to a firm reaction and authorizes the American administration to draw up, if necessary, lists of imported products and services subject to sanctions.

American industrialists rally around a strategy of acting pragmatically to seek new opportunities, notably in Europe. They have exerted pressure on the federal authorities with an eye to obtaining the opening of the EC's internal market.

2

The First Steps Towards a Community Policy

The origins of the Community Information Technology (IT) policy were in the early 1970s.

A communication from the Commission Council on November 21, 1973 examined the status of the European information technology industry and the trends in the evolution of its market. It noted that the market was increasing by 20% each year in Europe, but that 90% of the computers installed in Europe used American technology, with 60% of them supplied by IBM. It further pointed out the policy of "national champions" applied in Germany, France and in Great Britain, and emphasized that, in the field of mainframes, the four largest computing companies combined accounted for only 6% of the world market, a figure roughly equivalent to the weakest of IBM's American competitors. In the peripherals market, in spite of the relative success of some European companies such as Olivetti and Nixdorf, American domination was clear, as it was in the areas of electronic components and software. The Commission recommended the development of a European industrial foundation through cooperation between computing companies; it further recommended supporting it by an appropriate public open purchasing policy, and by the common development of applications for the benefit of users. It emphasised standardization, and the assembly and protection of personal data.

After reviewing this communication, the Council adopted a resolution on July 15, 1974, which defined the objectives of a Community computing policy and requested the Commission to submit proposals. The Commission put forward a series of proposals (set out in the following paragraphs), but it should be noted that until the end of the decade, the implementation of national policy remained principally the concern of individual governments. It was necessary to wait until 1981 for the launch of a true Community computing policy, with the ESPRIT program. Due to the ever tighter links between computing and tele-communications, it was 1984 before the first steps towards establishing a Community telecommunication policy could be taken.

The early Community moves in the field of Information Technology were modest. Their principal articulation was in a series of proposals by the Commission committing small amounts of Community funds to a variety of projects. Some were of a general nature, and others related to more specific applications. The Council, deliberating over the objectives and nature of which policies to follow in what had heretofore been considered as an industrial "sector", insisted on the creation of various committees, and replied to these proposals only after a long delay of three years. However, in time, a trend emerged in both the number of measures adopted and in the budgets allocated.

The first measure approved by the Council was the Resolution of July 15 1974. In order to give a Community direction to policies encouraging and promoting this sector (which up to that time had been treated solely at national level), the Resolution included a provision that the Commission would initiate a moderate number of projects of common European interest. These were in the domain of computer applications, the fields of standards and public purchasing, and also included industrial development projects having an element of trans-national cooperation.

This Resolution also recommended the establishment of a systematic Community program encouraging industrial R&D, computer applications and coordination of national promotional measures; along with these recommendations went the possibility of Community financial support. In pursuing this Resolution, the Commission presented proposals concerning common projects, studies and long term program.

2.1   THE PREMISES FOR COMMUNITY ACTION IN THE INFORMATION TECHNOLOGY DOMAIN

2.1.1   The first programs

On February 7 1975, the Commission proposed computer application projects in various fields: the creation of a data bank for the matching of human organs and blood types, based on national data banks whose information was put at the disposal of participating hospitals; a juridicial document retrieval system accessible throughout the Community; the development of a computer-aided design system for the design of logic circuits; management systems for the constuction industry; a processing system for import and export data, and financial management for the common agricultural market, and computerized sytems for the management of air traffic control.

In its decision of July 22 1976, the Council ratified only the first three projects, which were financed for three years at a cost of 1.2 million ECUs (1). On the same date, the Council also created a consultative committee for common computing projects.

Pursuant to a Commission proposal on September 10 1975, two years later (on September 27 1977), the Council adopted six decisions:

—three projects on the portability of software (that is, about the capacity of software to operate as far as possible on all types of computer), including a study of software languages, a study on conversion tools, and a study on the possibility of developing a common software interface for mini-computers;
—three studies on the subject of operational support in computing, including one on security and confidentiality of data, another on programming techniques, and a third on the establishment of data bases;
—an applied experimental project on the subject of rapid data transmission using the transmission of experimental data from CERN (Centre d'Etudes et de Recherches Nucléaires de Genève) to other European research centres using the OTS satellite;
—a set of exploratory studies in this sector intended to prepare a proposal for a medium term program.
—a study of the computer systems for the processing of import and export document data, and of the financial control of Common Market agricultural organizations.

The Council also confirmed the role of the consultative Committee (which had already been put in place by the 1974 measure) in assisting the Commission in the management of the projects it had authorized.

The projects ranged in duration from one to four years and were financed by a total budget of 2 977 million ECUs.

Pursuant to a proposal of October 27 1976, (and this time three years later), on September 11 1979, the Council decided to end a long term project in the computer domain which had already been underway for four years, consisting of both general activities and industrial promotion.

At the same time, the Council set up a consultative Committee for coordinating and managing computing development programs, and established the rules for the management of activities concerning

---

(1) 1 ECU ≃ $1.25; £0.65; 6.8FF; 2.05 DM; 1540 lire;1.8 SF; 2.3 guilders; 133 Ps; 42 BF; 7.8 Kr; 0.77 Irish punt.

the "Promotion" section of the program. The consultative Committee was also given managerial responsibility for this area. Funding was assured by a grant of 25 million ECUs from the Community budget, of which 10 million ECUs were allotted to general operations and 15 million ECUs to promotion.

On the same date, the Council adopted a resolution concerning community action to promote microelectronic technology, in which it invited the Commission to propose concrete community projects in the areas of microelectronic technology (equipment manufacture and advanced methods), the training of engineers and technicians, and coordinated computer-assisted design and test systems. A portion of the 10 million ECUs allocated to the general operations in the long-term program were set aside for these new proposals.

The contents of the first part concerning general operations are as follows.

—Standardization policy: the designation of priority areas, the Community contribution to international standards, the application of standards agreed at the Community level by member States, and the dissemination of information.

—Public markets: methods of rapidly applying standards in public markets, common principles in setting up schedules of conditions, methods of assessing submissions, exchange of technical experience and identification of topics to projects of common interest for public purchasing bodies.

—General aspects of computing policy: collaboration in research and development by coordinating the activities of research centers, defining new research activities, by the sharing of both results and personnel between research teams.

—Medium-term computing studies bearing on the status and evolutionary path of computing in the Community, and providing the analyses needed to put together future proposals for Community action in the field of microtechnology.

—Effects of computers on employment and their impact on society.

—Confidentiality and security of data.

—Legal protection of computer programs.

The promotional aspects concerned general software development, specifically with an eye towards enhancing the portability of programs, as well as the development of new trans-national applications, or those applications having an impact on standards.

It can be seen that this list contains in embrionic (and tentative) form the bulk of those activities which would later comprise Community policy in the domain of Information Technology, the primary exception being voluntary research efforts.

Despite their modest dimensions, these measures have brought some positive results:

—they constituted a means of implanting the mechanisms which were later to be put into effect in a large scale in future programs. These included cooperative procedures for the selection of contractors and the awarding of contracts by industrial research centers, and universities; methods associated with budgetary management, the training of managers and management programs, evaluation of results from completed operations, and the outlining of strategic directions in future programs;
—work in specific areas has accomplished the following;

- the first stage in coordinating data banks for human organs and blood types;
- the harmonization of the juridicial system (LEXIS), which has resulted in industrial firms marketing software implementing this harmonization;
- a highly important contribution towards the development of real-time programming language (ADA);
- the initiation of the CADDIA project (Cooperation in Automation for Imports/Exports and the Financial Management of the Agricultural Market), which now allows exchange of customs and common agricultural market data;
- the creation of computerized systems for exchanging data between the main European ports, as well as between many of Chambers of Commerce. These systems are now operational;
- the building up of a significant experience base in high-speed data exchange protocols using satellite transmission channels.

In the period for which the long-term program ran (between 1976 and 1983), while gaining experience in launching of management systems for running of cooperative Community programs, a significant body of results was also achieved. After 1980, the Commission also initiated its first plan of action in the micro-electronics field, and started activities in the telecommunications sector.

2.1.2   The transition towards the main Community programs

Based on a Commission proposal of September 1 1980, the Council adopted on December 7 1981 a measure concerning Community actions in the micro-electronic technology field.

This measure was supported with a 40 million ECUs budget and was to have a four year duration. The program, whose objectives were revised in 1983 in order to align them with ongoing technical advances, concerned advanced lithography equipment, chemical and physical processing of silicon and III-V materials, the fabrication, packaging and testing of integrated circuits, and computer assisted design of very large scale integrated circuits.

The evaluation of this program's execution had yielded abundant information regarding both individual projects, and operational aspects of the program. The evaluation showed that, apart from one or two exceptions, the projects had achieved their aims, and had reduced the technology gap between Europe and the United States and Japan. Six new companies had been formed in order to capitalize on the results obtained. The biggest success however was to confirm that collaboratively executed trans-national research and development programs executed in collaboration can succeed in the European environment. The ESPRIT program, which the Commission had started to prepare in 1981 with heads of industry in the Information Technology field, was destined to follow on from these efforts. In setting its sights on a global strategy, the Commission capitalized on the success and the experience gained to overcome the last pockets of political resistance. Such resistance had up to then opposed the formation of a Community R&D capability of the dimensions that would be needed to face up to the growing challenges.

During 1981 and 1982, five fields were identified in which intra-European cooperation could yield the most benefit. These were: advanced information processing, micro-electronics, software technology and development, administration, and computer systems of integrated production. Following the initiative of Vice President Etienne Davignon, the managing directors of the 12 largest industrial groups in the Information Technology sector, and more than 100 high-level engineers were brought together in five "panels" and put together what were to become the contents of the ESPRIT program.

In order to put these initiatives into effect as quickly as possible, and to test the organizational structures of full scale cooperation while waiting for the definition of the main program, on December 12 1982, the Commission proposed at the Council the launch of a pilot program lasting one year, supported by a budget of 11.5 million ECUs. This involved 15 projects. Industrialists verified that each of the projects was ready to start immediately and that each one had a value in its own right in its impact on standardization. These projects dealt with in micro-electronics and functional analysis of administrative requirements.

The Council gave its agreement, and the pilot phase was able to commence on the date requested by the Commission. The experience gained, and in particular the relationships formed between heads of industry, academics and Commission officials would allow the preparation of the main ESPRIT program in some six months. This program was presented to the Council in May 1983. The ESPRIT program covered the five areas described above, and was now given a budget of 750 million ECUs, to be charged to the Community (which represented the start of an overall program committment of one and a half billion ECUs) over a five year period. However, the program was not adopted at this time due to the difficulties encountered by the Community regarding the British contribution. This was finally unanimously resolved and ESPRIT was adopted in 1985.

2.2   THE FIRST COMMUNITY ACTIONS ON TELECOMMUNICATIONS MATTERS

2.2.1   The initial moves

During a meeting held on December 15 1977, the Council of Ministers raised certain questions concerning Posts and Telecommunications. It stated its intention to examine the respective roles of monopolies and the private sector with regard to telecommunications, as well as to coordinate network development projects in the Community, paying particular attention to developments based on Information Technology.

In April 1978, the Commission set up a task force composed of experts from the member States on "Future Telecommunication Networks". This group was directed to produce a report on "the principles of harmonization applicable to integrated digital networks". This report, which was completed in October 1979, analyzed the practice of technical harmonization adhered to by the individual Government administrations and by CEPT (European Posts and Telecommunications Conference). In particular, it underlined the difficulties that efforts towards harmonization encountered. It recommended a harmonization process focused on the functional characteristics of networks and services, and in particular, the progressive integration of the evolving analog capacities in the national networks. It also recommended intensified efforts within CEPT. Taking note of the report of this task force, and also wanting to counteract the fragmentation of the Community's telecommunications terminal and network equipment markets which was harming the development of European industry in this sector, the Commission submitted three proposals to the Council on September 1 1980, concerning:

—the implementation of harmonization procedures in the telecommunications field;
—the creation of a community market for telecommunication terminals;
—the first phase of the opening of public telecommunications markets.

The Council's debates on these recommendations led to a stalemate in October 1982, due to differences of opinion between two member States concerning how to accomodate the interests of the community industry in this sector. It became necessary to wait for the launching of the global telecommunications action program at the end of 1983 and beginning of 1984, before these matters could be resolved. However, certain measures had already been initiated by the Commission without delay.

2.2.2 The first measures in the field of telecommunications: prompt actions

2.2.2.1 The EURONET/DIANE information system

In June 1971, the Council of Ministers adopted a resolution aimed at coordinating the activities of member States in the areas of scientific and technical information and documentation (IDST) through the establishment of a committee composed of representatives of the member States and from the Commission. The initiatives it took sought to accomplish three objectives.

—The provision of a set of information resources and services in scientific and technical fields, capable of meeting the needs of corporate executives, researchers, administrators and teachers.
—The establishment of a physical network using modern technologies to transfer information from one point to another in the Community, and thereby enhancing the long-term economic viability of the system.
—The creation of an infrastructure composed of a physical network, an assembled data base, as well as appropriate techniques for inputting/outputting, processing and retrieving information. This implies information operations, and the training of potential users.

After a preparatory phase which lasted nine months due to psychological and political barriers to cooperative action within the member States and the PTTs, the system became operational in 1980.

In December 1975, a contract for the installation of a physical network (EURONET) was signed with the French PTT (representing nine Telecommunications Administrations in the Community). EURONET is a distributed network using the packet-switching technology of the French TRANSPAC X25 network. This network was completed in June 1977 with the help of the SESA companies, who carried out the work based on the TRANSPAC technology in France, and LOGICA, who worked on the British EPSS system.

During the same period, the DIANE (Direct Access Network for Europe) service, uniting all the information services available to the customer on the EURONET network, was progressively put in place.

In 1980, EURONET included four switching nodes, (London, Frankfurt, Paris and Rome), and nine access points (at each of these nodes, as well as Amsterdam, Brussels, Copenhagen, Dublin, and Luxembourg). For security reasons, the configuration of links allowed alternative routes to be set up between any two nodes. The transmission speeds were 64 kbits/sec between the nodes, and 9.6 kbits/sec between the multiplexers and the nodes. Access to Euronet could be gained via the telephone network, rented lines or the national packet transmission networks. A tariff system was also introduced whose charges were independent of the transmission distance.

In March 1980, 16 information services, representing some 90 data bases, were operational on Euronet. This number rose to 50 (approximately 500 data bases) in 1983. The system was able to support 2000 virtual access connections; this figure has now increased to more than 10 000 per hour.

The operation, based on a cost-sharing arrangement received 40 million FF (1) of financing from Community funds. The contract covered the costs of developing the network, as well as the writing-off of the initial deficit, which ceased in 1983. That year the Administrations resumed responsibility for the network in their capacities as public operators.

The DIANE system continues to develop, and today consists of more than 200 information service providers in Europe offering approximately 1000 data bases and realising a turnover of nearly two billion ECUs in 1990.

2.2.2.2 Pilot programs in the field of administrative and/or commercial data communication

In the early 1980s, the Commission launched pilot programs in administrative and commercial data communication. These programs permitted a comparison between theory and actual practice,

---

(1) 1 FF ≈ $0.18.

norms and their application, and between services offered and the needs of users.

The programs also allowed certain users including the Commission itself, as well as other Community institutions to play the role of "leading edge user". Indeed, by the targetting of these pilot programs, the Commission stimulated both development, and electronic data exchange among Community institutions, and between the Commission and the member States, national administrations and private firms.

There were three such pilot programs: the first two, CADDIA (Cooperation in Automation of Data and Documentation for Imports/ Exports and the Management of Financial Control of the Agricultural Market) and INSIS (Inter-Institutional Integrated Services Information System), were launched in 1982. The third, TEDIS (Trade Electronic Data Interchange Systems), was introduced later, in 1987.

● *INSIS*

The INSIS program sought to promote communication between the member States and Community institutions. It ensured coordinated and harmonized operations using new technologies combining data and text processing, and using advanced telecommunications systems. The program was approved by decision 82/869/EEC of the Council in December 1982. A Consultative Committee of users, composed of representatives of member States and Community institutions, determined the requirements and prepared proposals designed to develop integrated information systems.

The priorities of this program were:

—the establishment of electronic document transmission and electronic mail between the administrations of member States and the Community institutions, in order to reduce the delay in delivering these documents,
—the creation of video-conferencing systems between Community institutions and the member States.
—the establishment of a simple and convenient access system for Community information, the greatest part of which is stored in data banks run by the Commission.

The INSIS program led to the launching of a first generation of electronic messaging. This is beginning to be used operationally by the Commission services, and will ultimately allow communication with other Community institutions.

Within the framework of the INSIS program, experimental video-conferencing studios have been built in Brussels and Luxembourg.

These studios are widely used by the Commission services and during the parliamentary sessions. Indeed, electronic document transfers concerning parliamentary questions have already taken place between the Commission and the European Parliament. The technical infrastructure for electronic communications, with permanent display installations, is now being installed and will be operational shortly.

An enormous quantity of information is stored in the Commission computers: statistical data, judicial documents, archival materials, etc. In addition, the member States possess similar data bases, and several commercial firms also offer services in this field. In order to facilitate the access to these data bases, the INSIS program provides the support needed to set up access infrastructure for Community data bases, as well as to develop data bases of inter-institutional interest.

The INSIS program, along with the CADDIA and TEDIS programs (described below), will probably take on greater importance, with the amalgamation of information from each Administration and the need of some of them to "oversee" the establishment of the single internal market. The latter will herald the free movement of people, goods and services, and capital. This new freedom will necessitate the communication of more and more data which will require advanced computing and telecommunications. This is what the Commission calls "the European nervous system": and these programs will now be called upon to meet this need.

● *CADDIA*

CADDIA (originating from a preliminary study carried out pursuant to the second program of priority actions in 1977, and the subsequent Council decision 82/607/EEC of July 28 1982, approved by the Council on March 26 1985), was set up to coordinate activities of the member States and the Commission. It concerned a long-term program to investigate the use of telecommunication techniques in the Community information systems. Specific applications include imports and exports, financial management of agricultural organizations in the Market, and the gathering and dissemination of statistical data on Community trade.

The program's objective is to satisfy the urgent needs in the field of electronic data transfer in the customs, agricultural and statistics areas in a rapid and reliable manner for the realization of the unified Market of 1992 .

This program has a Steering Committee made up of representatives of the member States and Commission leaders in the relevant sectors. This Steering Committee has approved a long-term development program, and on June 1 1987, the Council extended the duration of the program through 1992.

The program's main operations relating to customs consist of the following; improving the customs tariff data base TARIC (Tarif Intégré Communautaire); specifying a message system for the transmission of TARIC codes; specifying computerized error detection methods; progress in the development of the customs messaging system EDIFACT (Electronic Data Interchange for Administration Commerce and Transport); initiating a pilot program administering the quotas in the SPG system (Système de Préférences Généralisées) and finally, completing the first phase of the SCENT project (System Customs Enforcement Network) which facilitates the electronic transmission of urgent messages for the anti-fraud campaign.

In the agricultural sector, the AMIS system (Agricultural Market Intelligence System) is widely used by the authorities in the day-to-day management of Market organizations. With the extended application of IDES (Interactive Data Entry System), a reduction in the number of telexes transmitted and the elimination of data retransmission can be expected.

The FIS (Fast Information System) project has become operational and offers an "electronic newspaper" for the publication of agricultural information.

In the area of statistics, emphasis has been given to the introduction of norms in statistical applications, the standardization of statistical reports, and the establishment of an information gathering system.

In the field of external trade statistics, the accent is placed on the improvement of world trade data, the creation of a data base on SPG imports, and improved access to tariff data bases.

Agricultural statistics operations concentrate mainly on the EUROFARM project (a data base on agricultural structures).

In addition, within the scope of CADDIA, several feasibility studies are in progress with the goal of setting up a national server organization in each member country in order to bring about the transfer of customs, agricultural, and statistics data between member States and the Commission in a coordinated and uniform manner, and conforming to international standards. This significantly contributes to the realization of the "European Nervous System" linking the Administrations referred to above.

The CADDIA program has been the subject of an evaluation study designed to define and adapt its objectives and activities in the perspective of the creation of the single Market by the end of 1992.

● *TEDIS*

On October 5 1987 (decision 87/499/EEC), the Council of Ministers adopted a Community program concerning the electronic transfer of trade data using communication networks: TEDIS (Trade Electronic Data Interchange Systems). On April 5 1989, it consented to the association of certain independent countries, notably members of EFTA (European Free Trade Association), with the TEDIS program.

The TEDIS program began on January 1 1988 with a budget of 5.3 million ECUs and a planned duration of two years.

The TEDIS program has essentially two objectives:

—the coordination of work, mainly concerning standards and security, for the development of electronic transfer systems for trade data,
—the creation of awareness in potential users and European equipment and software producers.

TEDIS has coordinated a certain number of electronic data transfer projects (EDI: Electronic Data Interchange) launched under the intitiatives of various European industrial sectors, including:

ODETTE	in the automobile industry
CEFIC-EDI	in the chemical industry
EDIFICE	in the computer and electronics industries
EAN-COM	in the distribution and retail sectors
RINET	in the insurance sector

TEDIS has provided logistical support to these industrial sectors and has encouraged information transfer between them.

These sectors had common concerns at the time that EDI (Electronic Data Interchange) was set up, consequently TEDIS has been designed to deal with inter- as well as intra-sector issues.

As well as coordinating the pan-European projects in each sector, the first phase of the TEDIS program has actively contributed to standardizing the interchange of EDI messages by supporting the work of the EDIFACT Board for Western Europe. This support has facilitated the definition of the EDI user needs concerning telecommunications matters. It has also assisted the search for appropriate solutions to judicial questions raised by EDI, the investigation of how the security issues of EDI messages can be resolved in a dynamic and open environment, and also the establishment of procedures and structures needed to ensure that messages and software conform to the international EDIFACT standards.

The launching of the EDIFACT program has coincided with a growth in interest in EDI: TEDIS has therefore been able to actively contribute to the development of EDI in Europe, and consequently has been able to respond to needs as they have arisen. Activities in the period 1988–1989 further highlighted the need for action to address future concerns.

A second phase of TEDIS will be launched during 1991, and its prime objective will be the general application of EDI in the European Community until 1995.

2.2.2.3   *The first moves in favour of competition in the telecommunications arena*

The provisions concerning the "Treaty on Competition", and the judicial precedents established in this field by the Court of Justice have general application to the telecommunications sector. Before competition was specifically addressed in Community telecommunications policy, the treatment of certain topical "cases" by the Commission and the Court had defined the principles subsequently taken up in the context of common policy.

Two principles appeared to be fundamental.

—In the case of a State monopoly or a public enterprise (including the telecommunications administrations) which contravened the rules on competition due to national legislation or government instructions, an action against the member States in violation can be instigated under articles 37 and 90, paragraphs 1 and 3.

—Companies in the fields of telecommunications or information technology cannot engage in agreements or pursue policies that violate the rules of competition. It could concern public network operators (telecommunications administrations) or "companies" in the sense of articles 85 and 86 when these operators not only carry out government instructions, but also act as independent business entrepreneurs.

● *An example relating to the telex service: the "British Telecom" case*
This case constitutes an example of the above principles and has been noted as such.

The circumstances arose from a claim registered by a private British message transmission agency against the relevant British telecommunications authorities. The claim concerned the restrictions imposed by the General Post Office and, after the Telecommunications Act came into force in 1981, by British Telecom. It concerned the carrying of telex traffic between independent countries (for example between the continent of Europe and North America) and the transmission, in telex or telefax, form of messages received on computer links.

A decision from the Commission on December 10 1982 (JO L 360, December 21 1982, p.36, 12th report on Competition (1982), point 94) condemned these restrictions as an abuse of a dominant position. The Italian government contested this decision and took the matter to the European Court of Justice. The British government supported the Commission's defence before the Court (Case 41/83).

In its judgement handed down on March 20 1985, the Court rejected the action brought by the Italian Republic. This judgement was reported in the 15th report on Competition (1985), points 95–101, in the following terms:

—"By the ruling of March 20, 1985, The Court of Justice has rejected totally the action of the Italian Republic against the Commission aimed at annulling the decision against British Telecom (BT)."

—"In its decision, the Commission had considered that British Telecom—which is a legally constituted public limited company, and holds a legal monopoly over the control of telecommunications systems in the United Kingdom—had exploited its position in a manner constituting an abuse under the terms of Article 86. This abuse of a dominant position resided mainly in the fact that British Telecom had prohibited private message-forwarding agencies in the United Kingdom from transmitting messages by means of the British Network, when the source and destination were in other countries. Several rules laid down by British Telecom between 1975 and 1981, which governed the use of the public telecommunications network by private subscribers, in fact included such restrictions."

—"The Commission considered the restrictions imposed by British Telecom to be abuses, for the following reasons:
- they prevented British message-forwarding agencies from providing a new service to their customers established in other member States in the Community,
- they subjected the use of public telecommunications installations to conditions which were neither technically nor commercially justified,
- they imposed a competitive disadvantage on private agencies compared with the national agencies and authorities of other member States."

"... In this case, the Court concluded—just as the Commission had before it—that British Telecom, in its action against private forwarding agencies, had not acted in the public interest, but in its capacity as a business. The operations by which British Telecom runs its public telecommunications installations and puts them at the dispostion of users in return for payment of charges, indeed constitutes a business activity. This applies also to the promulgation of contentious rules. The power conferred on British Telecom to make these rules is limited to the fixing of tariffs, to other operating constraints, and to the conditions of services offered by British Telecom to its users..."

"...The Court went on to reject the assertion that the application of article 86 to the case in point would violate article 222 on the grounds that it would prejudice the right of member States to create or maintain public monopolies for certain economic activities. It noted that, although British Telecom was in possession of the legal monopoly to run its telecommunications networks and to put them at the disposition of users, it did not hold a monopoly on the supply of auxiliary services, such as that of the re-transmission of messages on behalf of other users..."

"...Moreover, making use of a new technology allowing a more rapid transmission of messages constituted technical progress and was accordingly in the public interest and could not be regarded, of itself, as an abuse."

"Contentious issues were also not covered by article 90, paragraph 2..."

"...The applicant has in no way established that the condemnation of a contentious rule by the Commission compromised, from an economic point of view, the achievement of British Telecom's specific mission. Even if the speedier transmission of messages allowed by technological progress leads to some decrease in the income of British Telecom, the operation of private forwarding agencies attracts towards the British public network a certain volume of international messages, and the accompanying income. It had not therefore been definitely established that the activities of these agencies had overall negative effects on British Telecom..."

The British Telecom ruling clearly confirmed that *the rules of competition in the Treaty apply to telecommunications administrations.*

The ruling also shows that the Court favoured a *narrow interpretation of the rights of monopoly.*

● *Matters relating to terminal equipment*
The Commission raised these matters in Germany, Belgium, Italy, the Netherlands, and Denmark. They concern the *illegality of the extension of the monopoly to the terminal equipment sector.*

In Germany, the following circumstances arose:

—The cordless telephone case, reported as follows in the *15th report on competition policy* (1985):

"...The Commission had addressed the Federal Republic of Germany who, through the intervention of the Bundespost, wished to extend the telecommunications monopoly to the area of cordless telephones."

"The German government had in effect intended to reserve for the Bundespost the right to supply virtually all the cordless phones to be connected to the public telephone network. Only instruments intended to be used in conjunction with a PABX supplied by an economically independent operator were to be excluded from this monopoly. The Commission considered that even the partial monopoly of the Bundespost fell within the scope of article 37, paragraph 1, line 2, of the EEC Treaty. In effect, instruments imported from other member States cannot be freely marketed in Germany, even if they conform to the standards in force in that country. After the intervention of the Commission, the Federal Republic of Germany abandoned its plans to extend its monopoly to cordless telephones."

The case of modems was also addressed in the 15th report on competition in the following terms:
"...The Commission also intervened with respect to the extension of the Bundespost monopoly to modems. This case concerned the connection of instruments which operated in digital mode to an analog telephone network. As before, the German government held the opinion that this case concerned equipment which formed an integral part of the telephone network and which, consequently, could only be supplied by the Bundespost services. The only exception to this rule was for modems used to link private digital instruments which made no use of the public telephone network."
"The Commission considered that in this case the matter came within the scope of article 37, paragraph 1, line 2, of the EEC Treaty, and that, in linking the provision of service that was constituted for the utilisation of telephone lines with the supply of modems, the Bundespost abused its dominant position under the terms of article 86, in so far as it was the owner of the network. In order to correct these violations of articles 37 and 86, the Commission made known to the German government its intention to implement a ruling under the terms of article 90, paragraph 3..."

In June 1986, the Commission reached an agreement with the Bundespost over measures designed to modify the legislation which established the Bundespost as the sole distributor and owner of modems in Germany.

With regard to Belgium, the Commission received a complaint that exclusive rights had been accorded to the Belgian Post and telecommunications authorities to import and supply low-speed modems and the first telex terminals at all locations. After being informed that the monopoly in this case was incompatible with the EEC Treaty because it did not allow equipment suppliers of other EEC countries direct access to the Belgian market, the Belgian government indicated to the Commission that it intended to reform these arrangements within three years.

A similar claim was expressed regarding a monopoly on the importation and sale of modems and the first telex terminals in Italy.

In this case, modems and the first telex terminals to be connected to the public network could only be supplied and installed by SIP (Società Centrale Servizi Telegrafici). This arrangement affected the imports of modems and telex terminals coming from other member States since manufacturers of this type of equipment did not have direct access to the Italian users.

After the Commission's interventions, the Italian government announced that it was going to reform these arrangements.

Finally, in the cases of Holland and Denmark, the Commission held an inquiry on the existence of exclusive rights on the importation and sale of terminal equipment in the two countries. The Dutch government announced that it was going to reform its arrangements, and negotiations are still in progress with the Danish authorities.

Having acted in a case-by-case manner, the Commission took global measures in matters concerning terminals and services, based on article 90.3 of the Treaty. These measures are presented in the third part of this chapter.

2.3   STANDARDIZATION IN INFORMATION AND TELECOMMUNICATIONS TECHNOLOGIES

● *Introduction*
Within the framework of the general policy developed by the Community for all sectors of its economy, standardization is seen as being a decisive factor in the Community's strategy for the development of information and telecommunications technologies. Common standards having an open form are indispensable in assuring the end-to-end inter-operability of terminal equipment, and for the interconnection of telecommunications networks.

The Community policy on standardization which is followed in the area of Information Technology is based on guidelines approved by the Council of Industry Ministers in May 1984. The following points form the framework of that policy:

—The priority to be given to international standardization (ISO, CEI, CCITT,CCIR), in particular in the area of communication between open systems (OSI).

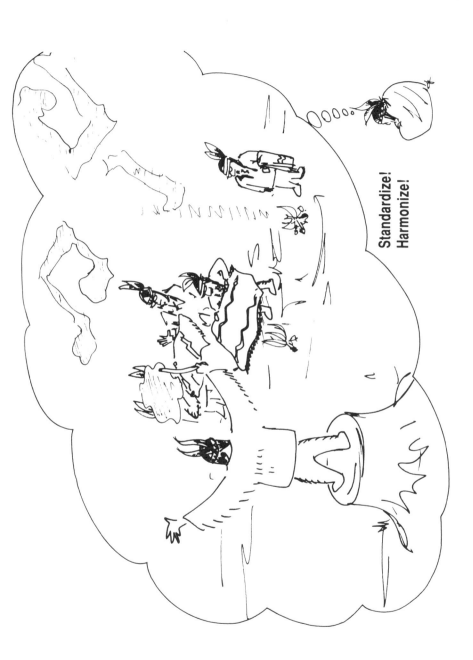

—The overriding necessity to monitor coordinated international standardization in order to ensure the interchange of data and the inter-functioning between systems. This implies a mechanism covering the following aspects:
- reference to international standardization in national systems of standardization;
- the refinement of standards, when these are not sufficiently precise to fulfil the requirement of data interchange functions;
- the development and implementation of means of testing, to ensure the conformity of equipment to these standards;
- reference to standards in public purchasing and regulatory activities within the member States.

—The systematic utilization of structures and procedures that are already available:
- the use of competent technical bodies: ESC, EESC, and now ETSI;
- application of the Directive 83/189/EEC, which imposes on member States the obligation to inform the Commission of their national standardization programs, and to supply the means of coordination in this field.
- the need to consider the convergence between the Information and Telecommunications technologies, to ensure coherence between the normalization efforts in the two areas.

● *The pre-eminence of international standardization*
The overriding importance of international standardization has been vigourously reaffirmed in the fields of information and telecommunications technology.

The achievements that have been accomplished at the international level represent an enormous investment of resources; the long and difficult work, such as the elaboration of the ISO standards for OSI, the ISDN (Integrated Systems Digital Network) recommendations made by the CCITT and the MHS (electronic messaging), should not be underestimated.

The fact that information exchanges and business trends must now be considered in the international context justifies opting for international standardization, which is inconsistent with isolated and fragmentary solutions.

● *The application of international standardization*
The harmonized application of international standards in the European context is one of the questions that has necessitated a more precise analysis, particularly in view of the difficulties encountered

in past years, when attempts were made to use the standards to ensure the interchange of information within the Community. The problems of simple file transfer between computers and the problems of data retrieval from different systems by different types of terminal, are well known. The problem of incompatibility has affected the field of telecommunications; services such as teletex, videotex, and mobile radio-telephony clearly illustrate this phenomenon.

Standardization in these areas must take into account the technological evolution of systems and their effect on the very nature of the standards. For a long time, standards stood for years, or even decades, through the application of "good practices". Now it has become necessary to specify the architecture of new systems and to rapidly draw up specifications which will allow such systems to communicate. This evolution therefore imposes constraints which are due to:

—*complexity:* standards are complex, and the systematic digitalization of all the networks requires that the binary (0 and 1) signal streams to be perfectly defined;
—*speed:* the pace of technological innovation demands that the specifications of standards to be achieved with minimal delays in order to avoid their obsolescence;
—*quality of data inter-change standards:* many standards have been set up to provide information to system designers. Too few standards are usable as a base for ensuring exchange of information because of the diversity of options that they contain.

In the field of telecommunications, inter-working requires the precise definition of:

—end-to-end compatibility;
—the connection to the network which should be harmonized with the level of detail needed for the free circulation of terminals in the Community and the access to information services on the basis of standards open to all.

All these restrictions have demonstrated the necessity for a sufficient refinement of international standards, at least for a transitional period, in order to avoid divergences and incompatibilities which call the utilization of international standardization into question. This refinement requires technical work which must be put in the hands of technically competent bodies.

Also, it is worth noting that, although certain misgivings about the risk of isolation resulting from the refinement of standards prevailed at the beginning of this exercise, the situation has since evolved rapidly. Heads of industry and those responsible for standardization in several countries have indicated a desire to see international standardization applied with a greater precision, and comparable initiatives are now all taking place in the different countries. This return of interest in solutions, which in fact correspond to common objectives, should lead to a preference for an improved convergence of views in favour of the setting up of more precise standards, even in the forum of international bodies.

● *The role of the Commission*
Since 1984, the Commission has taken initiatives needed to facilitate the expression of needs, the fixing of priorities, and the development of work programs. These were then entrusted to the relevant European standardization bodies working on information and telecommunications technology. It has been assisted in these tasks by committees; the "Senior Civil Servants Committee in Information Technology" (SOGITS), and the "Senior Civil Servants Committee in Telecommunications Technology" (SOGT). These committees are consulted about a wide range of problems concerning standardization. They give an opinion before making a committment to work on standardization, and play a central role in the various stages of the establishing Community policy. In addition, the procedures laid down in Directive 83/189 define two channels of information:

—national standards bodies inform each other of their standardization programs in order to determine whether the intentions of any one of them to carry out some work do not lead to the drawing up of a new European standard;
—member States inform the Commission of anticipated technical rules in the course of development.

SOGITS has specific technical authority in this field which allows it to call for detailed consultations, and it is consulted before any work is passed to the standards bodies by way of the committee of the Directive 83/189.

● *The role of standards bodies*
The technical harmonization work required for the smooth running of the Community market, and in particular, the elimination and prevention of technical obstacles, have traditionally been given to two European standards bodies CEN and the CENELEC. These bodies group together members' national standards institutes, and cover the countries of the European Free Trade Association (EFTA) as well as member States of the Community. This structure allows ample

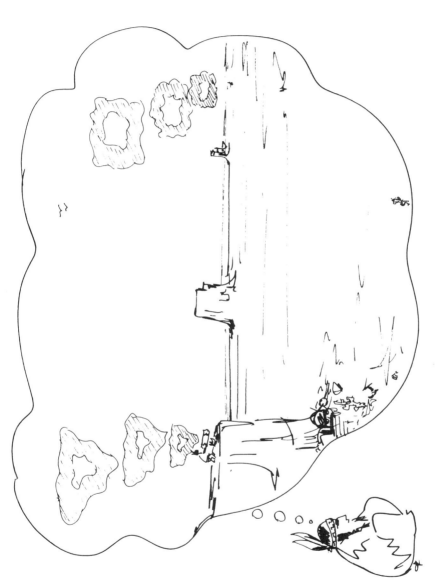

Different message protocols!

opportunity for consultation and participation for all interested parties through their national standards bodies. In addition, since the CEN and CENELEC are also members of the ISO (International Standards Organization) and of CEI (Comité Electrotechnique International), a large degree of consistency with work undertaken in the international context can be maintained. The statutory terms of reference of CEN and CENELEC allow them to develop European standards known as "EN standards", which are adopted according to precise rules following a public inquiry, and are used in national standards systems after adaptation, or at least following the removal of divergent technical specifications in existing standards.

These bodies can be mandated to carry out the required technical work, with the objective of developing an EN standard. Following the outline-contracts which were signed in 1985 between CEN/CENELEC, and between the EC and EFTA, the establishment of contracts for such work could include the grant of financial participation.

Since 1984, these bodies have agreed to undertake the necessary work in the Information Technology field and have progressed to laying down the structure and procedures appropriate to specific nature of this field.

The identification of common areas of standardization, notably in the overlap between the information technology and telecommunications sectors, has encouraged and justified the cooperation which has been established between CEN/CENELEC and CEPT. Since 1989, this has been extended to the European Telecommunications Standards Institute (ETSI), which has taken up, within the scope of its efficient organization, the standardization tasks of the CEPT. The standards prepared by ETSI and ETS are aligned with the practices of CEN/CENELEC, notably with respect to the taking up and application of technical specifications at national level (to be adopted on the basis of a carefully considered vote) thereby facilitating cooperation.

The primary goal in Community standards policy is to give these bodies the responsibility for ensuring that the dynamic development of standards corresponds to the role played by standards in a modern industrial society.

● *The role of industry*
In the area traditionally covered by industrial standardization, the role of industrialists is particularly important. First, industrialists have played a decisive role when the Commission reached the consultation stage within the context of the "Round Table" meetings.

From the end of 1983, twelve industrialists clearly indicated their interest in applying international standardization based on the OSI model (Open Systems Interconnection).

The participation of heads of industry in large measure shapes the provision of technical expertise, without which standardization work could not progress. Industrialists also have a decisive influence on the promotion of standards, and the creation of a "SPAG" type structure (Standards Promotion and Application Group) has allowed the consolidation of activities. Contacts between SPAG and similar organizations in the United States and Japan will allow increased progress in moves in the direction of harmonized worldwide application of international standards.

In the telecommunications field, most heads of industry in this sector cooperate within the framework of ETSI of which they are full members

● *The verification and certification of conformity to standards*
The means for verifying conformity are essential because they help to avoid deviation in their application if they are available at an early enough stage, and they give credibility to standards at the user level. The implementation of these measures requires that the methodology should be available, and that the laboratories and test centres are able to provide the relevant testing services.

The activities undertaken cover the telecommunications field as well as that of Information Technology.

● *Future evolution*
In addition, the Commission has presented the Council with a "global approach to the field of certification", which proposes the basis of a common system of operation for all economic sectors in the verification of the conformity to standards.

Finally, in the light of the experience gained, a global review has been undertaken of the functioning of the entire system of Community standardization. To confront the challenge of a smoothly operating internal market after 1993, significant improvements are needed, particularly with regard to accelerating the development, and the establishment of standards. This review will in all probability lead to a substantial revision of existing structures.

3

The Rise in Power of European Telecommunications

It is worth noting that, in 1984, the telecommunications companies (PTTs) of the different countries of the Community remained both powerful and protected, and that each country protected its national industry by following a policy of supporting national champions. As a consequence, the equipment and services market remained fragmented.

Two external factors began to have a significant impact: the political, industrial and commercial consequences of deregulation in America, and the initial steps towards a political concensus for concrete action to implement the internal market. This concensus would culminate with the signing of the Single Market Act in 1986.

3.1   EUROPEAN TELECOMMUNICATIONS IN 1984 —AN ASSESSMENT

3.1.1   Protected PTTs

With the exception of the United Kingdom (who, for general political reasons, had elected to follow a governmental "hands-off" approach, while reviewing the regulations with an eye towards opening the avenues to competition), for a long time the countries of the European Community sought to exist under the protection of their own monopolies. They were little disposed towards taking the risks (particularly in the social sphere) of radical structural changes that technological innovation was ultimately to force upon them.

This reticence to change found other justifications: an adherence to the principle of public service, which it was feared might be eroded by too draconian an application of liberalization; doubts about the rapid transformation in the demand for services; the force of habit; and a fear of becoming enmeshed in a system which could inevitably lead to a reduction in their own power.

The origin of European Telecommunications dates from the telegraph transmission networks. All countries sooner or later ended up by nationalizing their telephone networks for various reasons which were similar to those that, in the United States, had led to a private monopoly.

After a honeymoon period, in which numerous private companies co-existed, a confused situation developed where the networks multiplied, without being inter-connectable, leading to inefficient services.

The intervention of the State in this sphere became necessary in order to ensure that users as a whole were provided with satisfactory telephone service in terms of quality and price.

Thus on the eve of American deregulation, the situations in the European Telecommunications, although specific to the individual Community countries, had the following common characteristics:

—each country's postal and Telecommunications services were grouped under a single administration, the PTT;
—the PTTs had simultaneously the function of operating the network and services, and supplying and maintaining terminals connected to the network.

In 1983, the European market was the second largest in the world, with an annual growth of 5.3% ; it represented more than half the size of the American market, and about a quarter of the world market.

The EEC market, unlike that of the United States, was simply the sum of 12 autonomous national markets, supported by independent technologies and, at least for the largest countries, by national industries. These industries responded to their own needs, within the limitations imposed by the sizes of their own markets.

It was at this time that the American market, which was characterised by a network largely interconnected by the single operator (AT&T), became the focus of attention of the authorities. These authorities considered that control over the whole market by a single company (including the operation of all its lines, the fixing of its tariffs and standards, and the manufacture of equipment), was excessive. However, the European States, with the exception of the United Kingdom, continued to content themselves with much smaller markets. These states had networks which were badly connected to other national networks, were based on diverse technologies, and were established by relatively weak industries engaged in vigorous competition outside Europe. This fragmentation of policies and internal markets in the European Community emerges clearly from the following description of the status of each member State during the early 1980s.

Belgium represents one of the clearest examples of a public monopoly situation and its effects. The Telephones and Telegraphs Corporation (RTT) possessed, under governmental control, a monopoly of telephone and telegraph services by virtue of two laws passed in 1930.

This status was extensively applied, and encompassed the supply of equipment. The functions of network operation and regulatory authority were combined. The RTT operated under the direct control of the Ministry of Posts and Communications, on which it relied for definition of its investment programs as well as its purchasing policy, which were laid down within the framework of the government's industrial policy.

The situation for users was difficult. In addition, progress towards digitization was slow, tariffs were high, and delays in network connection could exceed six months.

However, the RTT has begun to relax its monopoly hold on terminal equipment provision, so that only the first telephone instrument and the first telex terminal had to be supplied by the company.

In **Denmark**, the State enjoyed a monopoly over public telecommunications, and had granted concessions to four companies in which it retained a majority interest.

In this way, four regional franchise companies co-existed alongside the PTT (Telecom Denmark); these were Copenhagen Telephone Company, Jutland Telephone company, Funen Telephone company, and South Jutland Telecom. The first two are SARL companies in which the State retains a measure of control, while the third is a municipal company and the fourth a public enterprise.

These regional companies had a monopoly over the supply of basic service within their own territory, while Telecom Denmark was responsible for national and international communications.

In reality, there was only one network in Denmark, which was without direct competition, and whose tariffs were fixed by the government. The PTTs carried out the regulatory function by delegation from the Ministry of Public Works.

The franchised companies also had a monopoly of equipment supply.

Value-added services were open to competition by private service providers in two-way communications.

The situation in **Spain** was characterised by the presence of a virtually total monopoly, accompanied by the absence of a single administrative body. The different networks were run by separate entities.

The telephone and packet-switching networks (Iberpac) were run by a public limited company, the Compania Telefonica Nacional de Espana (CNTE), whose capital was under the State control.

Since partial privatization in 1984, the Telefonica company has been only 46% controlled by the State. The latter is represented on the administrative council by a delegate who is usually the Director General of Telecommunications at the Ministry of Transports, Tourism and Communications. The government enjoyed certain special powers, notably the right of veto over certain strategic questions, as well as financial and tariff matters. The government reserved the right to renegotiate at any time the concession and schedule of conditions which were imposed on the company.

Telefonica was directly involved in the production and provision of equipment through its financial involvement in the ownership of different suppliers, to whom it gave preferential treatment in its purchasing policy. Since the approval of equipment was also managed from within Telefonica, the distinction between the different functions was extemely vague.

The telegraphic and telex networks came under the jurisdiction of the General Direction of the Posts and Telegraphs, who work under the Ministry of Transports, Tourism and Communications.

The market is not well developed, with a penetration rate that is still low. Ninety per cent of lines are linked to electro-mechanical exchanges.

The development of private data transmission networks was less advanced than in other countries, due to the absolute control exercised by Telefonica over the use of specialized lines and modems, as well as a restrictive approvals policy. There existed practically no value added services industry.

Tariffs were generally high, and even prohibitive in the short distance data line rentals sector (+ 140% compared with the European average).

In **France**, the Direction Générale des Télécommunications(DGT), which was under the Ministry of Posts and Telecommunications, exercised regulatory authority and monopoly over service, as well as the power to define the system of terminal equipment supply.

The Economics Ministry controlled the finance, service tariffs, and loans undertaken by the DGT and took part of its income to balance the national budget. In fact the DGT was the greatest source of cash flow in the French Administration at this time. It directly exercised, in the context of its monopoly, management control over the Telecommunications network in France. It also controlled services through the preferences of its subsidiary companies, of which it owned part or all of the shares.

The supply of terminal equipment was already quite liberalized. The DGT granted certification and always attempted to have available two supply sources for every item of equipment, including network equipment that it purchased directly.

France, which in 1974 was lowest ranked in Europe with regard to telephone density (behind Spain and Portugal), has successfully caught up, and now ranks sixth in the world (behind the USA, Sweden, Switzerland, Canada and Germany) with 94% of homes connected to the network. In the process, the DGT decided to exploit the new technologies and to install digital equipment directly into its network, which has made the French network among the most advanced in the world at present.

The DGT has also decided to operate specialized networks and services, such as:

—Teletel, which has placed France in the lead in tele-services. The PTTs have integrated, through the Minitel terminal, value added services through access to data bases;

—Transpac: a packet switched transmission network, serving more than 20 000 subscribers, with a growth of 60% in 1984. It has been considered as the leading network of its type in the world with a high quality of service;

—mobile car-phones: although growing rapidly (23% in 1984), the number of subscribers (10 700) remains limited due to the high price of equipment and the inadequacy of geographic network coverage.

Finally, France has shown itself capable of designing, implementing and operating space telecommunications with its Telecom 1 satellite launched on August 4 1984.

It should be noted that investments in telecommunications equipment are almost exclusively the responsibility of the PTT administration.

Great Britain is playing a pioneering role in Europe, having opened up its telecommunications to competition, with the 50% privatisation of British Telecom (BT) in 1984.

In fact from 1977 (and therefore before the American deregulation), Great Britain was questioning whether it should preserve the Telecommunications monopoly in its existing form (General Post Office), or separate the postal from the telecommunications service. On October 1 1981, this division was made, in the "British Communication Act".

The path towards liberalization opened in 1981 with the creation of an independent Telecommunications company—British Telecom— separated from the postal service. At the same time, the government introduced limited competition in the supply of equipment. In the following year, the government licenced Mercury Communications, another independent company, to operate a public network in competition with BT. During 1984, half of BT was sold on the London Stock Exchange.

In 1984, Mercury (a subsidiary of Cable and Wireless) offered its first international services from its own teleport, and started building a group of networks linking the most important British centres of activity.

In addition, the trend towards deregulation of a number of Telecommunications activities led BT to extend its activities to include professional value added services, with a view to opening them up to competition.

The direction taken led irreversibly to a shattering of the monopolies and widening the opening towards the private sector. The function of market regulation, which was totally distinct from the network operations, was put in the hands of OFTEL (Office of Telecommunications), which was created in 1981. OFTEL was set up particularly to ensure that BT and Mercury operated within the terms of their operating licences, and that they adhered to British standards as defined by the British Standards Institute. The approval and testing of equipment were carried out by the British Approval Board for Telecommunications.

In **Greece**, telecommunications were structured as a state monopoly exercised by the Greek Office of Telecommunications (OTE), a public company having a theoretically autonomous financial and administrative status.

All of GOT's activities were subject to government control, which was exercised by a large number of authorities, acting through a complex network of agencies. This control also influenced market regulation and GOT's investments, personnel policy, and tariffs. Investment policy for example, was formulated by both the Ministry of Transport and Telecommunications and the Ministry of the Economy. Purchasing policy with regard to procurement was formulated in the Ministry of Industry, and in the Ministry of the Economy for payments.

This status was consistent with an opening to competition in the terminal equipment area, but the network remained archaic, and the market was under-developed.

In **Ireland**, a law passed in 1983 (The Postal and Telecommunications Services Act) clearly changed the landscape of telecommunications, and allowed liberalization in the provision of terminal equipment and services.

The law separated telecommunications services from postal services and in 1984 set up a public company, Telecom Eiran. Although entirely owned by the state, this company operated in an autonomous manner and without government subsidy. It financed itself by its own means, albeit with an imposed upper limit on its public sector borrowing.

The monopoly applied exclusively to the public network, and ended at the customer's connection. Telecom Eiran had the authority to grant licences to private service providers, while excluding resale of capacity on rented circuits, or their interconnection with the public network. Aggrieved parties had the right to appeal to the Ministry of Communications.

The 1983 law also liberalized the supply of terminal equipment, with the exception of the provision of the first telephone instrument. The public company supplied an entire range of equipment in competition with the private sector. Its purchasing policy was open, and international bids for contracts were regularly invited. Technical approval was given by the Ministry of Telecommunications, which exercised the regulatory authority.

In **Italy**, the telecommunications structure was one of the most complex in Europe. Telecommunications were operated under a monopoly which the State could either exercise directly or grant concessions on; the telephone service was operated in part directly by the State, through the Azienda di Stato per i Servizi Telefonici (ASST), and in part by two franchise companies in which the State was the majority shareholder.

The ASST is an autonomous agency of the Ministry of the PTT and which had its own resources and assets. It was responsible for:

—the installation, maintenance and operation of the intercity co-axial cable telephone network;
—the operation of the intercity telephone service in 37 main cities;
—the operation of the international telephone service to European countries and countries in the Mediterranean basin.

The franchised companies were subsidiaries of the IRI/STET public group:

—The Societa Italiana per l"Escercizio Telefonico (SIP) operated the part of the national network not operated by the ASST: all urban networks, the remainder of the intercity networks, telegram-telephony, and automobile radio-telephone links. It was the largest company, and had exclusive relations with all its subscribers;
—ITALCABLE ran all international services with non-European countries, with the exception of the Mediterranean basin. It also ran the data transmission services.

The tariff policy was laid down by ministerial decree, independently of the costs and needs of the operators. This situation upset the financial equilibrium of the operators and distorted the investment policy. The SIP, which regularly ran at a loss, had heavy debts.

This situation was unfavourable to innovation. The South of Italy particularly suffered from a lack of infrastructure, even Telex was underdeveloped (51 600 subscribers in 1983), new services were not readily available, and the tariffs were high. It was only in 1984 that SIP obtained authorization to operate public switched data transmission networks, which resulted in a lag in this area also.

Luxembourg had a traditional type of PTT administration, "Administration des Postes et Telecommunications", which had a monopoly over all networks and services, and exercised the regulatory function jointly with the Ministry of Finances.

Nonetheless, by 1984, the user terminal equipment market had already been liberalized for about twenty years, with the exclusion of the first instrument.

The administration exercised approval of terminal equipment, but generally recognised approvals obtained in other member States without requiring a new technical examination.

In the **Netherlands**, Telecommunications have been controlled since the 1930s by the Dutch Posts and Telecommunications Administration (Staatsbedrijf der Posterijen, Télégraphie en Téléphonie).

The monopoly exercised by the Dutch PTTs affected regulation, the network, and its services. It was accompanied by a dynamic investment policy: the launching of network digitalization, and the installation of optical fiber networks in the four largest cities. In the early 1980s, in the light of the evolutions in America and Japan, a debate on the evolution of structures was begun. In 1982 an independent Commission (the Swarttouw Commission), set up on the request of the government in 1982, pronounced itself in favour of the liberalization and increased autonomy of equipment supply.

In 1984, the government instructed another Commission, the Steenbergen Commission, to consider plans for a new Telecommunications structure. In the following year (1985), it reached a conclusion favouring the transformation of the PTT into a SARL (Société à Responsabilitée Limitée—public limited company). The separation of network operator and regulatory functions, and the liberalization of the equipment and services market.

In **Portugal**, telecommunications was totally controlled by the State; three operating companies shared jointly in a monopoly over the network infrastructure, as well as that of basic and value added services.

—the company Telefones Lisboa e Porto (TLP), a public company, served Lisbon and Oporto, which represented 60% of the lines;

—the company Compania Portuguesa Radio Marconi (CPRM) handled most of the telephone and telex international links. The administrative council was named by the government;

—Correios Telecommunicacoes de Portugal (CTT), which corresponded to the PTT, provided all other services, and fulfilled the role of regulator, under the authority of the minister.

The right to operate certain specialized value added services has been conceded to specific users (banks, airline companies...). The market remains underdeveloped.

In **the Federal Republic of Germany**, the "Fundamental Law" of 1949 (the German Constitution) put Posts and Telecommunications in the hands of the federal government, thereby grouping together regulatory authority, infrastructure provision, and network operator.

The Deutsche Bundespost, a department of the Ministry of Telecommunications, exercised this monopoly while leaving open the possibility of opening up to competition certain services, as well as equipment supply.

In 1982, the function of equipment approval was detached from the Deutche Bundespost's prerogative powers and was put in the hands of the ZZF, a separate body based in the Ministry of Posts.

The law on Postal administration of 1955 established the organizational structure of the Bundespost: the postal and telecommunications services were to be jointly operated, the Bundespost was to balance its budget and could, if necessary, operate cross-subsidy between the monopoly services and competitive services. It enjoyed relative freedom in the setting of tariffs, with a veto power reserved for the "Postal Administration Council". It was not subject to taxation, but was required to remit 10% of its total annual revenues to the State.

The Bundespost had total control over the allocation of the frequency spectrum, as well as the standards that terminals were required to meet (including PABXs and equipment attached to them) which made the development of private local networks difficult.

The Bundespost exercised its monopoly over the network, public circuit switching, value added services involving voice transmission, main equipment connected to the line, and the maintenance of the private telex network.

The Bundespost was in competition with private enterprises for the provision of value added services for closed user groups, and terminals for certain public value added services (e.g. videotex).

Telex and public telephones were provided entirely by private companies.

Specialist data transmission networks have multiplied since 1975 (the date at which Bundespost decided to catch up in this area).

Bundespost was by far the largest customer in the German telecommunications equipment market. It devoted 815 million DM to a forward-looking satellite program which was not launched until 1989. The Bundespost was the major investor in Germany and its approval and certification procedures were considered to be the strictest in Europe.

Bundespost fell somewhat behind in the digital techniques field. It realised this in the 1980s, and throughout the decade decided to devote the necessary investments to the installation of wide-band optical fiber.

The debate in Germany over structural evolution started in 1984. It dealt mainly with the increasing overlap between telecommunications, computing, and the diverse applications of electronics. It also dealt with obstacles to the expansion of Bundespost's field of activity, and reconsideration of the subsidization of the loss-making postal services by the profitable telecommunications services, which limited investment in new services. It dealt also with the reviewing of the situation in the field of satellites.

3.1.2 "National champions"

It would seem from this country by country review that, at the beginning of the 1980s, European telecommunications administrations enjoyed special exclusive rights for the provision of services and operation of the networks. Accordingly, they dictated the choice of equipment used in the actual network, and give certification of approval of terminals.

In all these countries, however, internal debate is being generated as a consequence of developments in "telematics", that is the increasingly close connection between computing and telecommunications. This new discipline multiplies the number of possible services, and also raises questions over the wisdom of applying to the management of these services (whose infrastructure is only starting to be put in place) the complexity of the rules established to ensure the so-called basic services (essentially telephone and telex). These questions have come up at the time when the United States has modified the management rules of its own network (which, unlike Europe's is widely interconnected), and has partially removed Bell Telephone's monopoly power (operation, regulation, equipment supply, research and development).

Before the member States of Europe were to implement a policy of "controlled liberalization", under the impetus of the European Community and the Green Paper, they were to adopt two strategies in the industrial domain:

—Some would choose to expose their main national manufacturer to competition from a second, and possibly foreign, manufacturer, like the Netherlands who, in addition to Philips, gave approval to Ericsson; Belgium which is the home of ITT Europe, but is also supplied by GTE; Sweden, where the operator, so as not to be totally dominated by the powerful Ericsson, has created its own manufacturing subsidiary, Teli; Germany, who, despite its special relationship with Siemens, is also supplied by SEL, a subsidiary of ITT.

—Others were to gather around them a powerful national industry, like France, with the Alcatel-Thomson group, Great Britain, with the Plessey-GEC group and its "System X", and Italy with Stet-Telettra and its Proteo system.

This policy had certain disadvantages, which were increasingly revealed to be serious, as it became necessary to create new networks and to manufacture large complex switching systems.

The multiplicity of European companies relying on markets of limited size, and faced with non-European competitors, makes it impossible to obtain the necessary economies of scale, and does not allow the production of sophisticated equipment in optimal economic conditions.

This led to such strong intra-European competition that it neutralized itself inside Europe and aggravated the situation outside it.

It made the formation of competitive intra-European alliances difficult and did not encourage relationships with foreign companies.

In consequence, by 1984, no European company, except perhaps Siemens and Ericsson, had reached the minimum size necessary to face up to the competition of companies the size of AT&T or IBM.

3.1.3   The fragmented equipment and services markets

The European Telecommunications market remained divided and limited with regard to both equipment and services.

● *Equipment*

In 1989, the European Telecommunications equipment market represented around 25 billion ECUs which is about 20% of the world equipment market (120 billion ECUs). A comparison of the respective sizes of the North American (35% of the world market) and Japanese (11%) markets shows that Europe is in second place in the world market in this domain.

However, it is difficult to compare Europe with the United States and Japan. The diversity of standards among the 12 countries in the EEC constitutes an impediment to the necessary harmonization of equipment in the Community. Effectively, therefore, six different systems of digital switching exist in Europe (compared with three in North America and two in Japan), while the 130 million telephone subscribers in the EEC are connected to 15 public telephone networks by eight different types of telephone socket.

This wide variety of standards explains the continuing regional character of current telecommunications markets in the EEC. For too long, this has led to the creation of national quasi-monopolies formed by one or two exclusive suppliers to the public operators. The reservation of public orders for equipment for single national suppliers (representing 50% of the equipment market) has constituted a serious handicap for European business, and for the whole Community. For the European equipment manufacturing businesses, the small size of their national markets relative to the world market makes their research and development and production costs much higher, since the equipment cannot be readily used in other countries.

This partitioning has also compelled business users in the European Community to adapt to the standards and constraints existing in each country (equipment, terminals, tariffs). This adaptation has often meant that they have had to deal with 12 different operating companies if they are located in each of the EC countries (e.g. for banks and multinational companies).

The absence of a real single telecommunications equipment market therefore constitutes a handicap to both telecommunications suppliers and customers.

● *Services*

The European services market is worth about 85 billion ECUs (£60 billion, or $100 billion) which is about 22% of the world Telecommunications services market (400 billion ECUs).

Combined, the four most important countries, represent 83% of the European market in telecommunications services (Germany 26%, France 22%, Great Britain 21%, Italy 15%).

In order to attain an internal market, it is necessary to develop new services which are available to all users throughout the Community (mobile telephone, videotex, ISDN,...). This implies agreements on technical characteristics, the rate of development, and the rules of operational management.

The introduction of trans-European services should however be done in the context of adapting the regulatory structures, in order to take into account the changes in the nature of telecommunications services, and the global impact of the deregulation started in the United States in 1984, and in Japan in 1985.

The situation described above was to be given added prominence by two external events: first the effects of the American deregulation, and then the first steps towards the realization of an effective internal market.

3.2   THE FIRST SHOCK WAVES OF AMERICAN DEREGULATION

Deregulation in America, accompanied by the break-up of AT&T, brought two immediate consequences for Europe : a tariff war, and American pressure to rapidly penetrate the European market. This pressure was exerted simultaneously by firms and the Administration.

3.2.1   The tariff war

The tariff war between long-distance operators inside the United States brought about a general reduction in tariffs in North America.

The transatlantic market is worth about two billion dollars, if all services (telephone, telex, special communications, television, packet switched data, messages, etc.) are included. Individual users represent about 30% of this market, and businesses represent 70%, of which 20% is accounted for by the internal traffic of multinational companies. Great Britain accounts for approximately about 40% of this market, Germany 20%, France 11%, and Italy 8%. The key sector is therefore business traffic.

Against this background, AT&T, stripped of its local communications operations following its break-up, has naturally sought to rapidly increase its European traffic. In October 1984, MCI and GTE-Sprint, who were previously confined to the internal American long-distance sector, signed an agreement with British Telecom reducing the tariffs between the United States and Great Britain by up to 40%.

At the same time, MCI signed a contract with the Belgian PTT which allowed it, from a centre based in Belgium, to set up an electronic mail service reaching to all parts of the world to compete with the international telex and postal services. Belgium thus became a bridgehead in Europe for the distribution of electronic mail coming from the United States, as well as Brazil and Argentina.

AT&T immediately replied with a 5 to 29% discount on its international communications tariffs, forcing the international tariffs of the European PTTs to follow suit. The battle of tariffs therefore seemed to be waged at the expense of the European PTTs' traditional positions. An American advertising campaign even developed the theme "When you're in Europe, place your calls through the USA, it's much cheaper". Increasing competition developed due to both this situation and to specific tariff decisions made by Great Britain. For users in most other European countries, it was preferable to go via Great Britain to call the United States because almost all the countries applied higher tariffs for direct connection than for routing via London.

Even if such links were not always automatic, the risk of traffic diversion and the revenues involved had become considerable.

As a consequence, transatlantic tariffs tended to approximate the real cost of service provision. Each country wished to preserve, or even develop, its market share in order to compensate, by an increase in volume, for the reduction in tariffs. At the same time, decreases in costs were brought about by the installation of new transmission media, such as optical fiber submarine cable.

3.2.2   American pressures on Europe

The second immediate result of American deregulation was that, wishing to penetrate further the European market, and supported by the American Administration, American companies, launched a clever series of initiatives synergistically employing technical and liberal arguments. Most American industrialists consider the biggest barrier to be standards. They therefore advance the idea that a single European standard would be desirable, as it would then allow them easier entry into the European market. Of course, they also wanted this standard to be as close as possible to the American standards. This attractive argument is also astute in that American standards only "correspond" to international standards when the latter in practice align themselves with the American ones. In order to better promote this idea, certain American companies gave their common support to the argument that it is the market and demand which should determine standards, and that a universal set of standards is the key to a universal service.

It is interesting to note that in parallel, and inspired by other reasons (as indicated previously), this idea had also arisen within the European Commission at the same time. In fact, a "European" standard, which has the aim of ensuring the interoperability of services and the interconnectability of networks, would only make sense in those cases where an "international" standard did not exist, or contained ambiguities which allowed divergent interpretations, and thus created obstacles to interchange, particularly within the Common Market.

In order to support its industry, and in view of its balance of trade deficit, the American Administration decided to encourage openings in the market by bilateral discussions with different European countries. To this end, the U.S. Trade Administration launched a series of "MAFF" (Market Access Fact Finding) inquiries. These inquiries, which were quite aggressive in the case of some member States, provoked a reaction of solidarity at the Community level to such an extent that in June 1986 the Commission was able, with the general agreement of the member States, to organize a similar mission of government and industrial representatives in the United States.

The FCC instigated its own information search on the introduction of foreign companies into the American Telecommunications market, and advocated a retaliatory policy with respect to some of them. The main objective of this process was to prod the foreign governments into accelerating their telecommunication deregulation plans and to blunt the penetration of the American market by foreign firms.

After the passage of the Omnibus Trade Bill, which contained a section dedicated to telecommunications, the American Congress opened up a procedure allowing for retaliatory measures against certain foreign players judged by the Administration to be engaged in unfair competition. The USTR (United States Trade Representative) was to identify the European Community, along with South Korea (Japan having been dealt with separately) as being appropriate for negotiations designed to induce a greater opening of the market. The Community, for its part, did not accept such negotiations, but did not reject the possibility of pursuing discussions already in progress.

3.3 THE ACHIEVEMENT OF THE SINGLE MARKET

After the worldwide crises of the 1970s, the construction of the European Community had experienced a period of relative stagnation, with the notable exception of the creation and implementation of the European Monetary System (EMS). Budgetary quarrels impeded its progress until 1984.

At the end of this period, and on the conclusion of lengthy negotiations resulting in the entry of Spain and Portugal into the Community, it became essential to bring Europe back up to schedule. Consequently the Delors Commission chose a mobilizing objective and fixed a timetable for achieving it: "Objective 1992".

This objective—that of a single zone, common to 320 million people, cleared of the many obstacles to free exchange and cooperation among the twelve countries—had broad support among the economic and social spheres of influence, and was designed to allow the tackling of all those issues where concerted European action was recognised as a priority. The realization of a single internal market was to be effected by the implementation of the four great freedoms; the free circulation of people, goods, capital and services; technological cooperation, the environment, and the social dimension. The objective was announced in the European Parliament, and was put before the heads of State and government at the beginning of 1985.

A second phase of the relaunch of Europe arose soon afterwards. In order to make and to apply, within a fixed timescale, the 300 or so decisions which had been identified as necessary for the abolition of frontiers, practical changes allowing better decisions to be reached more quickly and more democratically had to be introduced. These were the objectives of the Single Act, which was adopted by the European Council of Luxembourg in December 1985, and then ratified by the national parliaments.

After this important reform of the Treaty, more than two thirds of the decisions concerning the establishment of the internal market could be taken by a majority decision in the Council of Ministers, instead of by unanimous consent. The European Parliament saw its powers strengthened. The indispensable cohesion of the Community could be assured by specific accompanying policies (in the regional, social and technological domains), ensuring that European construction would be beneficial for all regions, and social categories. The means of formulating these accompanying policies were established in February 1988, following three summit meetings, and the Twelve agreed to double the resources devoted to structural policies.

It is in this context that the equipment and services market—a vital element for the economy of the Community—and the associated Community legislation, were developed, primarily to guarantee that users' requirements, in all their diverse forms, would be met at the best possible price.

The establishment of this single market will facilitate the access to the market by industrial concerns and network operators from other countries. In order to preserve and strengthen their presence, and to acquire a strong position in the world market, European industrialists should constantly improve their competitiveness. The new dimension of the market in which they will operate, when cleared of its technical and commercial barriers, will provide a powerful boost towards this end. For their part, the Community institutions must be vigilant that effective opportunities for access to the markets of non-community countries are kept open, which are comparable to the opportunities in Community markets that are available to the industries of these countries.

—With these issues at stake, it is clear that 1984 represented a decisive time for European telecommunications. It is also clear, after this rapid review of the European telecommunications scene in 1984, that the reforms to be undertaken in this field would need to take into account external events, as well as the initial situations of Community partners. It would not have been very far-sighted simply to try to transpose into Europe the events that had taken place in the United States. From the start, the United States had benefitted from a vast homogeneous and vertically-integrated system: the Bell System, a single equipment supplier and a single network, which accounted for 35% of the world market. In Europe none of the national markets in the 12 member States is as large as any of the BOCs that were formed as a result of the breaking up of AT&T. Europe does not have the advantage of a homogeneous market like the one on which the American deregulation was built. The Community network is made up of a juxtaposition of 12 national networks having different technical characteristics. The march towards the realization of the single market will, however, modify in a positive manner the attitude of the member States towards Commission proposals geared towards constructing a Community policy for telecommunications.

This will demand a different approach from the Community to that followed in the United States. It will involve:

—the establishment of a "European Telecommunications Zone" by the development of interconnected networks and advanced services. This will particularly require the unification of standards, the mutual recognition of approvals, and the opening up of markets. It will also require the creation of a spirit of cooperation between the industrialists and network operators of the different countries, who have almost always acted in an independent, even competitive, manner;
—the definition of new rules based on common directions and principles in each of the countries in the Community, in order to be able to benefit from the potential for growth promised by network homogeneity;
—the adoption by the European States of common positions in external commercial policy matters when confronted with pressure or unilateral action of a protectionist nature.

An account of this common evolution of the 12 member States, under the impetus of the European Community, is presented in the following sections.

3.4 1984–1992: TOWARDS A GLOBAL EUROPEAN TELECOMMUNICATIONS POLICY

3.4.1 1984–1987: The program of action and the first political and regulatory decisions

It was at the end of 1983 that the Commission, which, as we have seen in the previous chapters, had already taken numerous, but scattered initiatives during the preceding years, launched the idea of one policy and one telecommunications zone. It presented a six-point program of action in November of that year at the Council of Industry Ministers, and proposed the establishment of a group of senior civil servants, presided over by the Council to assist in the detailed definition and execution of the program. This action program assigned three objectives to European telecommunications concerning the three major categories of players in the field:

—to provide *users* with the multiple advanced services on the best cost terms. These services would combine information and communications technologies to give a potential explosion of possibilities, and so contribute to supporting the competitiveness of the business community;

—to put at the disposal of the European telecommunications *industry* a large market capable of providing it with the economies of scale required by the high cost of developing the new equipment, and the need to confront relentless world competition.

—to allow the *network operators* to address the demand for advanced telecommunication services, to encourage them to make the decisions to invest, and to accept the corresponding risks in doing so.

● *The Commission proposed six specific categories of initiative*

—Cooperation between the network operators in order to install the new generations of networks and services in a European, and not merely national context: including Integrated Services Digital Networks (ISDN), integrated wide-band networks, or mobile radiotelephone networks and personal radio-messaging systems etc.

—Establishing a large European terminal and network equipment market. To this end, the action plan specifies standardization work to assure the interoperability of equipment, the mutual recognition of the certification of terminal equipment, the promotion of the use of advanced information services, and the opening up of national markets.

—Inducing network equipment manufacturing companies and network operators to cooperate in research and development relating to the integration of wide-band service networks. A common effort such as this is necessary to reach a consensus on common functional specifications, to reduce costs and risks, and to prepare industrial alliances.

—Allowing the less economically advanced member States of the Community to benefit from specific measures enabling them to get as much as possible out of technological progress and allowing them equal opportunities with the other member States to install new infrastructure and advanced services.

—Obtaining agreement from bodies of social opinion on the measures to be taken. Such agreement would take into account the foreseeable consequences on the structure and mission of the PTTs, as well as on the new qualifications that would be required in order that staff can easily operate the networks and advanced services.

—Avoiding an uncoordinated stance among member States in bilateral and multilateral negotiations with outside industrialized countries, and in the forums of the international organizations such as the ITU (International Telecommunications Union), the CCIR (International Consultative Committee for Radio-communications), the CCITT (International Consultative Committee for Telegraphs and Telephones) and GATT (General Agreement on Tariffs and Trade), and arranging that common positions be prepared in a timely manner.

These proposals were first presented in a succinct form on November 4 1983 at a Council of Industry Ministers meeting. At that time, the Council agreed to the Commission's request to make use of support from a group working in close collaboration with the Ministers of Telecommunications. This was the birth of SOGT (Senior Official Group for Telecommunications). Starting in November 1983, and throughout 1984, the Commission convened meetings of this group. On May 17 1984, the Commission presented the same plan of action to the Council (document COM (84) 277) in the form of a communication which was much more detailed, and was enriched by the observations of SOGT. This document was reviewed and discussed by the Directors General of Telecommunications on November 9 1984; then, together with a communication from the Vice President of the Commission, Etienne Davignon, by a meeting of the Council of Industry Ministers on December 17 1984.

These endeavours and discussions at different levels have permitted an initial Community telecommunications policy to be advanced pursuant to the guidelines laid down by the Commission. At its meeting of December 17 1984, the Council had, for the first time, recommended that the work in the telecommunications domain be pursued "on the basis of the main objectives" *in three directions:*

—the creation of a Community terminals and equipment market by a "policy aimed at the effective application of common standards throughout the Community established on the basis of international standards", and "the application, in stages, of processes for the mutual recognition of the certification of terminals".

—The achievement of better development of networks and advanced telecommunications networks by discussions on the realisation of infrastructure projects of common interest, the setting in motion of a program to develop the technologies necessary for the establishment of wide-band networks, and the definition and phased implementation of a video-communications system for the benefit of political leaders.

—The improvement of access to advanced telecommunications for the disadvantaged regions of the Community.

In doing this, the Council gave the political go-ahead for the work in hand and for the proposals presented jointly by the Commission and the Telecommunications Administrations.

In this way, *the Community telecommunications policy* was launched. The "Telecommunications" program of action gave rise to both concrete achievements and common initiatives only one year after its launch by the Commission. The latter benefitted from a remarkable spirit of cooperation on the part of the member States and the National Administrations.

One sometimes tends to forget the numerous decisions taken since 1984 in the six fields of action proposed by the Commission.

In July 1984, the Commission and CEPT signed a joint declaration of intent with regard to the creation of the market. Under the terms of the declaration, CEPT was given responsibility *on the basis of Community priorities*, to establish common standards as well as common specifications for the certification of terminals. An initial list of priorities was issued by the Commission and included in the CEPT work program. Top of the priority list featured (already!) ISDN standards, and those relating to the upper levels of the OSI protocol. A joint CEN–CENELEC and CEPT committee was created to avoid redundancy and to allocate responsibility between the two organizations.

On October 15 1984, the Council approved two recommendations that had been under negotiation since 1980. One anticipated the *a priori* harmonization of new services launched by the network operators. The other anticipated an experimental first phase of opening of their markets at the level of 10% for network equipment, and 100% for telecommunications computer terminals. In the same field, the Commission prepared a directive on an initial phase of mutual recognition of the certification of terminals (the phase of mutual recognition of conformance testing), which was transmitted to the Council in 1985, and adopted in July 1986.

With regard to the installation of Community networks, an initial demonstration of bilateral video-conference links between the Commission and five member States took place on November 15 1985, while a feasibility study was launched on the possibilities of extending these links to all the member States.

Preparatory work was begun on launching the RACE program (R&D in Advanced Communication Technologies in Europe), concerning the installation of wide-band integrated networks in the 1995–2000 timeframe. In addition, two studies were launched. One dealt with the concerted achievement in the field of second generation mobile telephony, and the other with the utilization of Community financial machinery to bring about the development of telecommunications in the disadvantaged regions of the Community. Finally, a sub-group of SOGT, the "Group Analysis and Provisions" (GAP), was created to establish common objectives in the development of telecommunications in the Community, taking account of the foreseeable evolutions in national situations. This group put under way three categories of work; on the ISDN, the broad-band network, and a second-generation mobile communications network. A timetable for proposals in each of these domains was fixed during 1985.

● *From the end of 1984 to 1987,* the Council adopted, on the basis of a proposal from the Commission, an impressive series of decisions which were to firmly consolidate the Community telecommunications policy.

—July 25 1985, a decision concerning the research and development program on advanced telecommunications technologies. This established a definition phase of one year's duration and was funded by a 20 million ECUs budget from the RACE Program.

—June 9 1986, a Resolution relating to the utilization of video-conferencing and video-phone techniques for inter-governmental applications, identifying specific areas in which the efforts of the Commission were to be concentrated after the first action phase which started in 1984.

—July 24 1986, a Directive on the first phase of the mutual recognition of certification of telecommunications terminals was adopted. It concerned the mutual recognition of results of terminal conformance tests to common certification specifications, which was to be put into effect by CEPT within the framework of the Declaration of the common intent signed in 1984. It announced a second phase, covering the

mutual recognition of certification of the terminals themselves, which the Commission presented to the Council in July 1989.
—October 27 1986, a Ruling instituting a Community program covering the development of disadvantaged regions of the Community by providing improved access to advanced telecommunications services. The ruling was based on the STAR program (Special Telecommunications Action for Regional development), and was provided with a budget of 780 million ECUs, for a duration of five years.
—November 3 1986, a Directive fixing the choice of common technical specifications to the set of standards for MAC-packets for the direct broadcast of television by satellite.
—December 22 1986, a Decision relating to standardization in the field of information and telecommunication technologies, as well as a Recommendation concerning the coordinated introduction of the integrated services digital network in the Community.
—June 25 1987, a Recommendation concerning the coordinated introduction of public pan-European cellular mobile digital communications in the Community, and a Directive concerning the frequency bands to be reserved for this application.
—October 5 1987, a Decision concerning the Community program for the electronic transfer of data for commercial applications using communications networks (TEDIS).
—December 14 1987, a ruling concerning Community activity in the telecommunications technology field: this refers to the main phase of the RACE program.

Events in 1986 and 1987 therefore consolidated the telecommunications policy launched in 1984 through dynamic action from the Commission, assisted by the member States and the national Administrations. In this period, the policy was ratified by the European Parliament, and this was followed by the approval of the Council. In June 1987, additional impetus was to given to this policy by the publication of the Green Paper on the development of telecommunications services and equipment in the Common Market.

3.4.2   1987–1988: A major step: The Green Paper and the Council Resolution of June 30 1988

In 1986, in parallel with the implementation of the action plan, the Commission began to prepare the Green Paper on a European policy combining liberalization and harmonization of the telecommunications market.

The Green Paper was published in 1987, and grew out of the belief that the time had come to initiate a Community-wide review of the fundamental institutional and regulatory adjustments to the European telecommunications sector which were needed in order to take account of the profound technical changes that were taking place in the global political and economic environment.

The general philosophy of the Green Paper can be summarized as follows: opening up the telecommunications sector to competition without destroying the organizational structures which maintain the integrity and viability of the infrastructures, and which allow the operators to carry out their public service functions.

- *The Commission had taken a number of factors into account when it researched common positions on the future regulation of the telecommunications sector*

—in the first place, the existence of the rules of the Treaty of Rome, particularly those concerning the free circulation of goods, the freedom to provide services, and the rules of competition;
—the external relations of the Community, especially with its main commercial partners: the EFTA countries, the United States, Japan, the Third World, and the obligations which flow from the GATT agreements,
—the impact of the new regulations on the industrial and commercial position of the Community;
—the progressive transformation of ways of thinking of member States with regard to new technologies, their social consequences, the conditions of their integration into private and professional life, and transitional measures relating to work.

- *Before making firm decisions, it was also necessary to take into account a number of requirements prior to a common move*

—the progressive character of the implementation of the modifications to be made, and as much consideration as possible of the particular circumstances of each State;
—the necessity to safeguard the role of the telecommunications administrations in the proposals for network infrastructures, allowing them to fulfil their public service obligations;
—the impossibility of defining a "natural" constant demarcation line between the "basic" services sector and that of "competitive" services, including, notably, "value added" services. Given technological evolution and the trend towards the integration of telecommunication services (whether they be in vocal, visual or written form), any definition can only be temporary and subject to review.

A consensus readily emerged among the member States to treat the voice telephone service as a basic service, reserved for the telecommunications Administrations in all member States. This service still represents 85% to 90% of current telecommunications revenues.

—A reduction in the delays entailed in the establishment and common application of international and European standards, needed to maintain the integrity of the network, and to make possible the interconnection of networks and services at the European and global scale.

—The necessary separation between the responsibility for establishing regulation, and the responsibility for network operation, the same players being ineligible from being both judge and judged at the same time.

—The entry of the multinational computer companies, as well as the Telecommunications Administrations into an open and competitive telecommunications market.

—The permanent control over possible abuse of dominant positions by either network operators or other "economic players" entering the telecommunications equipment and services market.

—the drawing closer together of the regulations of member States regarding access to the network for new competitive services, with particular attention to the points of connection to the network, tariff principles, the essential requirements concerning standards applicable to network equipment, and the allocation of frequencies,...

—the relationships between terrestrial telecommunications networks and satellite networks.

—the importance of external relations in the domain of telecommunications.

● *The recognition of these elements was translated into the following proposals in the Green Paper*
The common objective is the development of a strong telecommunications infrastructure and efficient services in the Community: the aim is to provide the European user with a large range of telecommunications services under the most advantageous conditions, to ensure their coherent development within the member States, and to create an environment open to competition, taking the momentum of technological development in progress fully into account.

There are ten proposals:

1) The maintenance of the principle of exclusivity or of special rights benefitting Telecommunications Administrations regarding the sale and operation of services on the network infrastructure. When a member State chooses a more liberal system for all or part of its net-

work, the short and long-term integrity of the general network infrastructure must be safeguarded.

The offering of bi-directional satellite communications will require a more detailed analysis later. The sale of these services should be authorized on a case by case basis as necessary in order to develop them on a European scale, without prejudicing the financial viability of the main service provider(s). A common conception and definition for the availability of the infrastructure should be achieved within the framework of Proposition 5 below.

2) The acceptance of the maintenance of the principle of exclusivity or of special rights to the benefitting the Telecommunications Administrations, for the provision of a limited number of reserved services, where this exclusivity is considered essential, at this stage, to safeguard their public service obligations.

The principle of exclusive service should be narrowly interpreted and be subject to periodic review, taking into account technological developments, notably the evolution towards a digital infrastructure. The definition of "reserved services" cannot extend the monopoly in a manner incompatible with the Treaty of Rome.

3) The open (unrestricted) offer of all other services, including "value added services" within the community. Network operators will themselves be authorized to offer such services, and common rules regarding access to the networks will be applied in a harmonized manner (see Proposition 5 below).

4) The necessity of applying common standards governing the network infrastructure and the services offered by Telecommunications Administrations or by service providers of a comparable size, in order to create or safeguard inter-connectivity on the European scale. These requirements should refer specifically to the Directives 83/189/CCC and 86/361/EEC, Decision 87/95/EEC and Recommendation 86/659/EEC.

The member States of the Community should effect and promote the sale of efficient communications on the European and global scale by the Telecommunications Administrations. This applies particularly to those services (reserved and non-reserved) whose provision at the European level has been recommended, such as the services defined in Recommendation 86/659/EEC.

5) The clear definition, within the spirit of the Community Directives, of the conditions of network utilization imposed by the Telecommunications Administrations, which are to be identical for all competing service providers.

These conditions relate to standards, allocation of frequencies, and tariff principles. The set of common conditions must allow the offer of an open network to users and service providers (Open Network Provision—"ONP").

6) The free availability of terminal equipment, with a reservation concerning approval and certification consistent with the provisions in the Treaty of Rome and existing directives. The provision of the first conventional telephone instrument could be temporarily excluded from the competitive bid.

Receive-only earth stations (ROES) for downward links from satellites should be incorporated into the treatment of terminals, and be subject only to certification procedures.

7) The separation of regulatory from operational activities of Telecommunications Administrations. Regulatory activities specifically concern the granting of licences, control over certification and interface specifications, allocating frequencies, and general monitoring of the conditions of network use.

8) The strict and continuous adherence to the rules of competition (articles 85, 86, and 90 of the EEC Treaty) in the activities of the Telecommunications Administrations, applying particularly to the subsidization of activities in the competitive sector and production services.

9) The strict and continuous adherence by the private service providers in sectors open to competition to the terms of articles 85 and 86, in order to avoid abuse of dominant positions.

10) The application of common Community commercial policy to telecommunications. Notification by the Telecommunications Administrations, of all agreements reached between them and outside countries, which could have an impact on competition within the Community under the provision of regulation 17/62. Communication of information, to the degree required by the Community to draw up a coherent Community position in the GATT negotiations and relations with other countries.

The discussion of the Green Paper took place during the second half of 1987, with representatives of industry, unions, and customers participating along with SOGT and the Directors General of the PTTs. Many written contributions were received from a wide range of organizations and firms, especially from the "United States Trade Representative" and associations representing American interests. The Commission studied the opinions it received, and on February 9 1988, sent a communication to the Council of Ministers drawing conclusions from its consultations and proposing to the Council an action program with a timetable for the progressive opening of the Common Market to competition, and the strengthening of European competitiveness.

At its meeting of June 30 1988, the Council of Telecommunications Ministers adopted a Resolution on the basis of this communication, which established in a more complete manner than previously the overiding objectives of a telecommunications policy for the Community. Its main themes were:

—to create or ensure the integrity of a Community-wide network founded on the inter-connectivity of all the public networks concerned;

—to progressively create a Common Market of telecommunications services, while taking into consideration the rules of competition set out in the Treaty, and defining, through the Directives of the Council, the conditions needed for supplying a network open for service providers to offer non-reserved telecommunications services;

—to encourage the creation of services at the European level, by acting in the areas of standardization, definition of the common tariff principles and in the encouragement of cooperation between network operators and others parties;

—to develop an open market for terminal equipment, based on the total mutual recognition of certification of terminals conforming to common specifications, to be established as soon as possible;

—to develop a Common Market:

- by separating regulatory and operating functions in telecommunication services,
- by applying the Treaty's rules on competition to Administrations and private telecommunications enterprises,
- by installing a transparent fiscal environment,
- by moving towards the complete opening of telecommunications supply and operating markets;

—to pursue the improvement in the functioning of European standardization, taking into account of the specific characteristics of this sector;

—to strengthen the cooperation of European industries in the fields of research and development;

—to create a social environment meeting the Treaty's objectives of improving the economic and social situation, by pursuing a dialog between social partners and creating the conditions necessary for the development of a social consensus;

—to completely integrate the more disadvantaged regions of the Community into the Common Market, consistent with the objectives of the STAR program;

—to define a common position on satellite communications;

—to take full account of the external impact measures taken by the Community in the telecommunications field.

The Resolution invited the Commission to put forward steps to be taken in order to attain these objectives.

● *A supplement to the Green Paper: the Green Paper on satellite communications*
The Green Paper of June 1987 had not addressed satellite telecommunication services. To fill this gap, the Commission adopted another Green Paper on November 14 1990, which was intended to launch a debate on the liberalization of these services.

Satellite communications constitute a domain where the Community market is particularly fragmented, and which is governed by regulations dating back a decade, while technology has advanced considerably in this period. The Green Paper on satellite communications puts four major modifications to the present situation up for discussion:

—the total liberalization of the terrestrial segment, by subjecting to a certification procedure those terminals that are limited to reception and that are connected to the terrestrial public network, while subjecting those that act as transmitters as well as receivers, to a certification and licensing procedure;
—free access to the space transmission capacity, under a licencing procedure;
—commercial freedom for suppliers in the space sector, implying direct marketing to customers;
—harmonization measures necessary to facilitate the supply and utilization of services at the European level (mutual recognition of licences and certificates, coordination of frequencies).

Discussions with interested parties are presently taking place.

3.4.3   The key elements of "The European Telecommunications Zone"

In the course of implementing Community telecommunications policy as defined by the proposals of the Commission and the resulting decisions of the Council, various degrees of progress have been achieved in each area. The most spectacular progress concerns the changes brought about in the monopoly structures of the PTTs; and the least satisfactory, so far, concern the setting up of advanced trans-European network services.

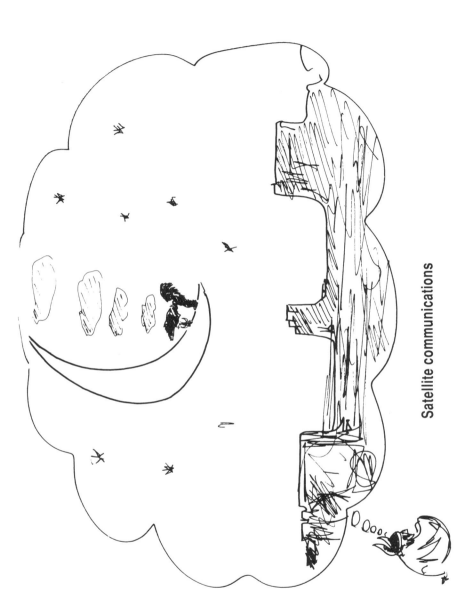

Satellite communications

● *The difficulty of establishing trans-European networks and services*
In an article which appeared in the Financial Times on February 24
1990, a Harvard University professor, F.M. Shere, wrote: "although
united politically, the United States did not become a real common
market until the railway network was largely in place and communications were improved, first by the telegraph, then with the telephone".
In a Europe where services play a growing and increasingly important
part of economic activity, the realization of a trans-European network
of advanced communications services will be a determining factor in
the political and economic integration of the Community.

In this spirit, since 1984, the Commission has attempted to give
an impetus to the creation of "trans-European" advanced service networks, with varying degrees of success for each transmission mode.
This action has influenced the integrated services digital networks
(ISDN), mobile communications, and broadband communication
networks.

● *The Integrated Services Digital Network (ISDN)*
Resulting from the new possibilities opened up by digitalization, the
IDSN constitutes an improvement in the present network by allowing the integration of narrow band services (most services except
those involving moving pictures) and giving the user the possibility
of access to a relatively large range of new services in return for a
reasonable investment.

The first efforts concerning ISDN, particularly in standardization,
started at a national level in the 1970s. It was not until December 22
1986 that, pursuant to a Commission proposal, the Council of
Ministers adopted a Recommendation on the subject. This Recommendation fixed a timetable for the establishment of common technical
standards, specified the nature of priority services and a timetable
for their introduction, and defined the level of penetration and geographic coverage. It was agreed that the provision of new services
(including facsimile, teletex, packet switching, audioconferencing,
call transfer, cordless telephones, and videophones) should be
achieved in three phases ending in 1993. The Commission directed
CEPT, and then in 1988, ETSI, to establish the standards for these
services based upon the CCITT recommendations, and to follow a
prioritized program and timetable. It should however be recognized
that it will be the tariff level and the structure of the ISDN that will
largely determine how quickly it can be introduced.

To assist the development of ISDN, a system to continuously eval-
uate the progress towards its introduction was put in place. The
governments of the member States must inform the Commission at

the end of each year of the measures taken to introduce ISDN, and of the problems slowing down its development. The European Parliament has also asked that an annual report on the progress in ISDN be prepared for it.

Three of these reports have already been presented, the first in 1988, and the second at the end of 1989. The first reported a delay in the implementation of the Recommendation of the Council of 1986, and divergence between the specifications applied by the first countries to offer ISDN services. The contents of the Recommendation were re-examined and, in July 1989, the Council adopted a Resolution articulating general objectives for the introduction of ISDN and asking the operators to sign an agreement amongst themselves. Two months later the operators signed a protocol agreement within the framework of CEPT, which committed them to the process of introducing a set of ISDN networks. The second report showed an improvement in the coordination of ISDN systems, but its evaluation found that the present situation was not yet satisfactory. In the third report, prepared at the end of 1991, further progress was recorded, although convergence of specifications had not yet been achieved.

Industry was encouraged to participate in this process of coordination as well as to manufacture equipment conforming to the standards recommended by ETSI, based on standards recommended by CCITT. When ISDN becomes an interconnected network at the European level, (which is not yet the case), it will contribute strongly to the realization of the single internal market, and will allow European industries to strengthen their position on the world market.

● *Mobile cellular telephony*
Mobile communications, as studies carried out in this area have shown, are poised to undergo considerable development during the next decade. First among these services is mobile cellular telephony, which allows the user access to telephone communication wherever he goes.

At present, mobile telephony is only available in certain limited regions of the Community, and is based on specifications that are not universally compatible. This means that a user with a mobile telephone attached to his vehicle, for example, will only be able to use it in certain parts of his country, and not at all beyond its frontiers. This situation is totally unsatisfactory from the point of view of the formation of a universal Community economic zone, and the Commission wanted to take advantage of the introduction of second generation mobile telephone systems in order to set up a single Community system that can be used in all the member States.

CEPT drew up specifications for a second generation digital cellular mobile telephony system, the GSM system (Special Mobile Communications System), and an agreement was reached between the Telecommunications Administrations to allocate the frequency bands 905—915 MHz and 950—960 MHz to the new services.

In order for this system to have a real chance of becoming the single trans-European system needed by the Community to face a strong growth in demand, it was imperative to take measures enabling it to be set up with the shortest possible delay, consistent with the time needed to prepare specifications and the time needed by industry to gear up to supply equipment.

The Commission therefore proposed a Recommendation to the Council on the coordinated introduction of pan-European public digital mobile communications, and a Directive reserving the necessary frequency bands for these services. The Council adopted these two measures in June 1987.

The Recommendation provides for a mobile cellular digital telephony system, on the basis of the GSM technical specification prepared by the CEPT, to be introduced in the Community and which started in 1991; a timetable was fixed, with planned objectives for the penetration of the service. The Directive reserved the common frequencies to be allocated throughout the Community for this service, to the extent justified by use.

These measures should allow a rapid and ordered transition from the present incompatible fragmented mobile telephony system to the single unified system which will be progressively available throughout the Community.

● *The radio paging system*

The radio paging system constitutes another type of mobile communication service experiencing a growing demand. This system uses mobile terminals of reduced size and cost to deliver an alarm signal or a short alphanumeric message to the pager, thereby prompting the user to contact the calling party by means of the conventional telephone network.

As in the case of the mobile telephone service, several incompatible types of radio paging service exist in the Community. This situation has led the Commission to act with regard to this service in the same way as for the mobile cellular telephone system.

On a proposal from the Commission, the Council adopted a Recommendation in November 1990 for the introduction of a pan-European public radio paging system starting in 1992, based on the ERMES common standard put into effect by CEPT, and fixing a

Mobile communications

timetable with penetration targets for the system. At the same time, the Council adopted a Directive reserving frequencies for the progressive introduction of this service.

● *Cordless phones and Telepoint*
Cordless telephones, which first appeared at the end of the 1970s, constitute another mobile communication system for which the Commission put forward a proposal for a Recommendation and a Directive in order to promote the establishment of a trans-European system.

The situation in the Community in this field is also characterized by the existence of incompatible national standards. Here, however, the mere use of instruments, mostly coming from South East Asia and marketed without conforming to any European standards, can involve risks as far as the utilization of frequencies is concerned.

In the 1970s, the CEPT had set up a CT1 standard, which was not introduced throughout the European Community because it led to an equipment design which was too costly to compete with the above-mentioned illegal imports: France and Great Britain, for example, retained their own national standards.

Starting in 1985, in the face of growing demand in this area, Great Britain developed a national digital cordless telephone standard known as UKCT2. CEPT also (though not so promptly) set up a standard in the same field called DECT (Digital European Cordless Telephone). These two digital standards, though incompatible, allowed three types of service to be offered with the same specifications:

—a conventional cordless telephone for residential use;
—the Telepoint system, made up of termination nodes installed in public places and along roads, by means of which users could make calls (but could not be called);
—mobile links, within business premises, served by a PABX.

The customer can use the same portable terminal for these three types of service. This terminal is already available at a lower cost than a mobile telephone terminal, which makes its use attractive, even with the inherent restrictions with regard to reception of calls. In addition, the DECT system has a more advanced design, and is better equipped for use in conjunction with a business PABX.

The UKCT2 standard, which has existed since 1987 (whereas the DECT standard has only been available since 1990), has been applied in the United Kingdom, where the government has granted Telepoint operating licences to four companies. This system is also of interest to other countries—including France—which has been able to install it since 1990. However, having been developed outside of all Community standardization, it would be difficult for UKCT2 to be widely accepted and to be adopted as the Community standard. It is necessary to wait for the start up of the DECT system to have available a truly trans-European system offering the high traffic capacities desirable for Telepoint application and links involving business PABXs.

These cases illustrate the difficult situations that can arise when a member State justifies the launching of a service based on urgent demand when the Community standard is not yet ready. The former then creates systems based on incompatible standards, and it is finally the user who suffers the consequences of this incompatibility. To avoid these problems, the Commission, working from the results of strategic studies in the field of new service development has drawn up a schedule for the establishment of standards. This schedule is aimed at making them available as they are needed, while supporting in every way possible the work of the standardization bodies (in particular those of ETSI) in order to reduce the delays in adopting these common standards.

● *The anticipation of future networks while bringing about necessary transitions: IBC (Integrated Broadband Communications)*
The IBC should allow the co-existence on one network of all services (including the transmission of moving images), that is, it should provide the infrastructure and service functions for any existing combination of services, from narrow band to broadband with a view to their functional integration. If the services presently intended for businesses are different from those intended for individual use, the differences between these categories of user will partially blur with the introduction of new services. The use of Teletel in France has already pointed the way towards this evolution.

It has been predicted that by the year 2000, the great majority of jobs will be in the service sector, and that a large portion of people working in this sector will be able to work from their homes. This will mean that domestic work-stations will need access to the same range of services as currently required by businesses. Advances in switching and transmission techniques, as well as the emergence of intelligent networks, will allow these services to be easily combined and mixed. In the long term, the communication needs of businesses and individual users will converge. It appears to be essential from now on to adopt an IBC strategy combining the needs of businesses and individual users at the start, thereby also achieving economies of scale. The strategy should also take into account high-

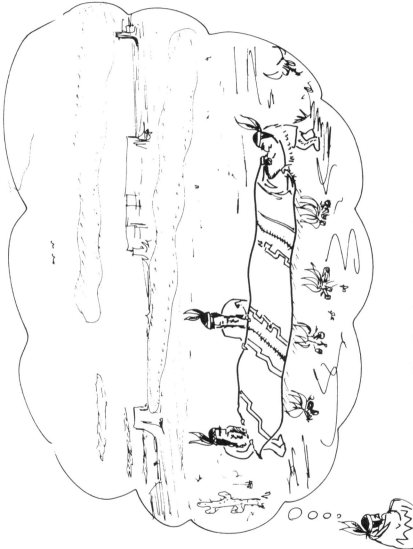

Broadband communications

capacity needs in information transport of the large electronic equipment users, such as large research centres, the chemical and pharmaceutical industries, manufacturing (automobile and aerospace industries) etc.

The reduction in the cost of transmission which we are now experiencing is certainly a desirable objective, but it would appear to be even more important to functionally integrate services in such a way as to be able to offer value added services in those forms best suited to the user. One of the most attractive characteristics of the IBC environment is that it will give service providers and users the ability to integrate different services in a flexible, or even ad hoc, manner. For example, this would allow a user to add a data service to the existing voice transmission service, to temporarily increase the communications capability in order to set up a video-conference, or to integrate computing, telecommunications and audio-visual services. It is necessary for Europe to develop an advanced telecommunications system (IBC) capable of forming a foundation for the emergence of new economic activities, thereby creating new opportunities for employment and for the promotion of international competition.

The RACE (Research and Development in Advanced Communications in Europe) program has the goal of making an active contribution to the development of IBC systems, by organizing coordinated and cooperative effort among the players involved, and by using the catalytic effect of research work.

The RACE program had been prepared during the course of 1984 within the context of close cooperation with the network operators, industrialists in the sector, research laboratories, and users.

The implementation of the RACE program started with a preparatory phase (held from July 1985 to December 1986) that was supported with a budget of 22 million ECUs. Its objective was to precisely define the contents of a large R&D program to meet the RACE targets. These initial investigations allowed the IBC to be studied as a reference model, and evaluated the most critical technological areas from the standpoint of the research effort to be undertaken.

On the basis of the results of this preparatory phase, a proposal for the main program was submitted to the Council in September 1986, and approved by the latter in December 1987. The Council granted the program a total 5-year budget, of 1 100 million ECUs, of which half would come from the Community budget.

The RACE program is not solely concerned with R&D; one of its other important objectives is the achievement of a consensus among all participants in the sector on future technologies and a strategy for the introduction of IBC. It has three main components:

—the quest for consensus on development and implementation strategies for the IBC, and the functional definition of systems;
—technological research and development designed to provide European industry with the advanced techniques necessary for the realization of the IBC;
—functional integration allowing the evaluation of service functions, in particular by the creation of pilot applications.

The program was planned with the following stages:

—mid-1988: first hypotheses on the configuration and environment of the IBC,
—mid-1989: decisions regarding a first IBC network and a strategy for its introduction,
—mid-1990: infrastructure projects of tested and validated systems,
—mid-1991: agreement on the architecture and common functional specifications.

The program, now being carried out according to the plan, issued a first call for submissions at the beginning of 1988, which resulted in the launching of 46 projects; a second call in July 1988 allowed 40 more to be started. These projects put to work 90 consortia, involving 306 organizations. These included 130 small and medium-sized businesses, 84 research institutions and 15 Telecommunications Administrations. Among these organizations, 27 were based in EFTA countries.

The follow up to the program in progress will mainly concern IBC applications. The third program framework for Research and Development, adopted by the Council on December 7 1989, was funded by a budget of 489 million ECUs over a five year period. The fields covered included intelligent networks, mobile technologies, image systems, information security, and service engineering. Full scale demonstrations are planned.

While the RACE program is being carried out, other R&D activities have been taking place in the field of telecommunications applications. Among the programs are AIM (Advanced Informatics in Medicine), DELTA (Development of European Learning by Technological Advance), and DRIVE (Dedicated Road Infrastructure for Vehicle Safety in Europe), which respectively concern the fields of medicine, education and transport. These three programs, which are still in the pilot stage, are designed to assure real-time interactive-mode data and image communication for public service purposes. They respond to new needs and create synergy among the participants, thereby exploiting the advantages of the European dimension. These programs are bound to have positive consequences for European electronics and telecommunications industries.

In addition to the pursuit of these three programs, the third program framework plans the preliminary research and development work needed for the establishment of a network of European data communication services between those national and Community administrations responsible for the progress of the single market. This particularly concerns the social security, customs, indirect taxation, and border police administrations. These networks should form the framework of a European nervous system for the communication of adminstrative data.

The program framework has an anticipated budget of 380 million ECUs for this group of activities.

3.4.4   The components of the equipment and services sectors of the single market

3.4.4.1   Liberalization and harmonization

Anticipating the final decisions of the Council on the implementation of the Green Paper, in 1987 the member States put national legislation in place, while the European Community established or proposed Directives.

● *At the Community level, since 1987, legislative measures have been taken or are in the process of being adopted. The most important are described below.*
They concern:

—the terminal equipment market;
—the opening of network operation markets;
—the services market;
—the protection of personal data.

Other important initiatives have been taken recently by the EC in the realm of external relations, and these will be briefly listed here.

—*The liberalization of the terminal equipment market*   The Commission launched three initiatives to open the terminal equipment market to the entire European Community, and to harmonize the conditions of approvals, by making use (depending on the objective) of articles 100-A or 90-3 (1) of the Treaty. In order to open the market for these types of equipment, it was appropriate to not only take away the monopoly over their sale from the PTTs, but also to arrange for approvals received from each member State to be transparent and mutually recognized by the other States.

The procedures for harmonization and mutual recognition of approvals in force in each of the member States were upheld in order to save manufacturers and suppliers the time and expense that parallel procedures for the same type of equipment would normally have entailed.

A first stage towards this goal was achieved on April 24 1986. The Council adopted a Directive, proposed by the Commission, establishing an agreement for the mutual recognition of conformance tests on any given type of terminal equipment in a member State in accordance with common technical specifications.

Once a certificate of conformity was granted by a member State on the basis of such a test, the other member States were no longer authorized to insist on new tests for equipment that had been so certified.

Nevertheless, a manufacturer who wished to sell his product in the Community was still required to follow the administrative procedures of each of the other member States in order to obtain approval of his product.

This situation will change as soon as the Directive on Legislative Convergence of the member States concerning telecommunications terminal equipment, proposed by the Commission on August 17 1989, and adopted by the Council on April 29 1991, takes effect in the member States—by November 1992 at the latest. This Directive will have the effect of regulating both the marketing of terminals, and their connection to the public networks.

The Directive allows the manufacturer to choose between two procedures for evaluating the conformity of his product to the harmonized standards. The manufacturer can opt either for "a CE type of examination" in which the product is subjected to tests in conformity with the relevant common technical rules in the laboratories of third-party bodies, or he can choose the "declaration of CE conformity" which calls on a certification body to rule on the basis of tests carried out by the manufacturer himself. If the manufacturer opts for the latter, more flexible procedure, he is asked to apply a quality control system and to agree to a "CE monitoring" system that provides for periodic

---

(1) Article 90-3 gives the Commission the opportunity, at its own discretion, to take measures needed to rectify violations of the rules of competition. Article 100-A is taken as the basis for the harmonization measures needed to establish the internal market, on the proposal of the Commission, within the framework of the process of cooperation between the Council and the European Parliament.

examination, as well as unannounced spot checks, in order to verify that the quality control system is being applied in the proper manner.

This procedure should allow manufacturers to produce terminal equipment in accordance with the harmonized standards established by the European Telecommunications Standardization Institute (ETSI) (which are transformed into mandatory technical rules by the Directive) and to offer this equipment for connection to the public network on the Community market without further formalities

Invoking article 90-3 of the Treaty on May 16 1988 the Commission adopted a Directive opening the terminal equipment market to competition. Although the majority of the member States had liberalized this market (as demonstrated by the analysis of the policies of different member States), for institutional reasons, some contested this Directive at the Court of Justice. However, a consensus now exists on the contents of the Directive. Moreover, in its ruling of March 19 1991, the Court in large measure validated the position taken by the Commission.

—The opening of the network operator market
These markets are often mistakenly referred to as "public markets". In reality this relates to procedures for awarding public works contracts used by network operators and by public and private telecommunications equipment suppliers who benefit from exclusive or special rights.

With regard to the awarding of public contracts in the telecommunications sector markets, the Council had already adopted (on November 12 1984) a recommendation anticipating that network operators would give the established suppliers in the member States of the Community the opportunity of submitting tenders for the supply of network equipment.

However, it must be recognized that, although this had the merit of stimulating national administrations to begin modifying their internal procedures and above all, their habits, this recommendation did not yield significant results. A report published by the Commission on October 11 1988, frankly admitted that "the results ... are disappointing in the number of public tender offers, at the level of response from suppliers in other member States, and even in the information provided by the member States to the Commission on the implementation of the recommendation".

In October 1988, these findings led the Commission to send the Council and the European Parliament a proposal for a Directive that would be less flexible than the earlier recommendation, and that would cover the procedures for awarding public contracts. The range of these contracts included areas which had up until then been excluded from the Directives in force concerning public contracts (sectors of telecommunications, energy, transport and water). The main object of this proposal was to ensure that businesses in the Community, regardless of their nationality, had an equal opportunity for access to the work and supply markets, or to services involving telecommunications software, generated by telecommunications organizations.

To achieve this, two principles were established: information from suppliers (publication of periodic updates on the trends of the market, advertising of tender offers and the results of previous contract awards), and the transparency of the procedures for awarding contracts (conditions to be met regarding delays in submission of bids, the selection of candidates or submitters, and objective, non-discriminatory criteria for awarding contracts, to be fixed in advance).

The Directive is aimed at public or private telecommunications bodies benefitting from exclusive or special rights to offer services on the telecommunications market. However, the Directive does not apply to contracts in those areas where other entities operate under objectively similar conditions: for example, the purchase of terminals for resale or leasing, or purchases relating to services which are open to competition. In addition, the Directive does not apply to the equipment supply markets, the provision of services where the cost exceeds 600 000 ECUs, or to installation work markets worth more than five million ECUs.

The technical specifications of the contracts should give priority to common European standards and specifications, except where this is not possible, in which case the national standards can be utilized.

The flexibility of the Directive derives from the fact that purchasing bodies can choose from among several procedures for awarding public contracts: this includes open call for tender; call for tenders from candidates selected or retained through a qualification procedure; a procedure negotiated with a call for competition; and procedures negotiated without reference to competition.

The recognition of the commercial interests of the Community with regard to other countries is achieved in the following manner: no guarantees given to offers coming from companies not established within the Community; the right to turn down an offer if more than half of its value derives from products originating outside the Community; preference given to equivalent offers associated with products originating within the Community, equivalence being established when the difference between their total values is less than 3%. The whole range of provisions of the Directive can be extended to offers from businesses in outside countries with whom the Community has made an agreement ensuring effective comparable access for Community businesses in these countries.

The Directive was adopted by the Council on September 17 1990. In addition, the Commission put forward a proposal for a new Directive on July 25 1990 concerning the monitoring of the opening of telecommunications sector markets, and is preparing a proposal on the awarding of contracts in the general services markets.

To accompany the latter Directive, a further one is being approved by the Council which will provide an effective mechanism of recourse to firms who consider themselves to have been unfairly treated in the awarding of contracts.

—Telecommunications services markets

The liberalization of almost all services, and the harmonization of the conditions of network access for service providers have been tackled simultaneously.

Three events have led to a modification of the conditions for the provision of telecommunications services within the Community: 1) the significant and rapid increase in the number of these services resulting from the progress of information technology; 2) the 1986 decision to launch the unified internal European Community market before the end of 1992 (which provides for the free movement of goods and services, as well as for respect for the rules of competition set out in the Treaty); and 3), the modification of the telecommunications regulations in other industrialized countries.

The retention of monopolies in the provision of telecommunications services would not have been compatible with any of the legal, political, or factual consequences of the above three events.

These reasons led the Commission to act. *Because it had always considered that measures for the liberalization of telecommunication services went hand in hand with the measures for the harmonization of these services*, it proceeded to prepare two Directives: one concerning the liberalization of services; the other, the ONP Directive, concerning their harmonization. By the end of 1988, the Commission itself adopted the first of these Directives, which related to the competitive conditions applicable to telecommunication services. Along with the Directive on terminal equipment, this Directive is based on Article 90-3 of the Treaty. In consideration of the deep reservations expressed by some member States on the contents of this Directive, and because of the links between it and the proposal for a Directive on conditions of network access for service providers (the ONP Directive), the Commission, while adopting the text, decided to postpone its implementation. The delay was intended to allow the adaptation of the Directive of the Commission to take into account viewpoints of member States. This was done in order to arrive at the widest possible consensus on its contents where possible, and also to obtain in a similar manner an agreement of the Council on the ONP Directive (founded on article 100-A of the Treaty).

On November 7 and December 7 1989, two meetings of the Council of Telecommunications Ministers of member States, under the presidency of the French Minister, M.Paul Quiles were devoted to the Commission's proposal for a Directive on competition in the services domain, and on the "ONP" Directive.

As a result of these two meetings, which had been preceded by numerous earlier meetings between the representatives of the National Administrations and officials from the Commission, a global agreement was reached on the contents of these two texts. However, a large majority of the member States continued to voice reservations over the utilization by the Commission of article 90-3 for the liberalization of services. This agreement married the liberalization of telecommunications services to the harmonization of the conditions of access, which must of necessity be combined.

Indeed, if liberalization is necessary for the development of new trans-European services (for the reasons indicated above), it is appropriate to avoid the inevitable malfunctioning which would ensue from the absence of harmonized regulation. Examples of effects to be avoided include the provision of services incompatible with the use of different equipment or network standards; tariff wars that would jeopardize the viability of networks and infrastructures; or a disorderly choice of frequencies, ruling out trans-European mobile communications services.

Adherence to the requirement of interoperability has remained an "optional" requirement in the Community legislation. It would only apply in "justified cases". It is obvious however that the availability of trans-European value added services throughout the Community necessitates adherence to this requirement which, in view of the interactions occurring in communications, means the precise observance of a detailed and unified specification or standard.

In summary, the balance between liberalization and harmonization was taken into account in the Council's compromise of December 7 1989, which on June 28 1990 implemented the two Directives whose contents will now be analyzed.

The Commission Directive concerning competition in the telecommunications services markets stipulates that member States should ensure the abolition of exclusive or special rights with regard to the provision of telecommunications services other than those of voice telephony. However, the Directive does not apply to telex, mobile radiotelephony, radio paging, or satellite communication services.

—That member States take the steps necessary to guarantee the right of any commercial operator to provide the telecommunications services to which the Directive applies. Those member States that subject those services to an authorization procedure or a declaration to ensure adherence to essential requirements, must take care that the authorizations are granted upon clear, objective and non-discriminatory criteria. Any ultimate refusal should be duly justified, and a procedure of appeal against such a refusal should exist.

—The member States were requested to communicate to the Commission before December 31 1990, the measures taken in conformity with the obligations of the Directive of the Commission. They were also asked to inform the Commission of any existing regulation, or of any pending plans to establish new authorization procedures, or to modify existing ones. Until December 30 1992, they may refuse commercial operators the right to offer simple resale of capacity in leased lines to the public .

—Packet or circuit switched data is the subject of specific arrangements.

—Plans for authorization procedures or declarations relating to these services were requested to be conveyed to the Commission not later than June 10 1992.

—The provision of such data packet or circuit switching services will be able to be subject to a set of conditions—forming a public service schedule of terms which should be objective, non-discriminatory and clear. These conditions will be able to deal with essential requirements, business regulations affecting the conditions of continuity, the availability, and quality of service. They will also deal with the measures aimed at safeguarding the general economic mission given to a telecommunications body for the provision of this service, in those cases where the actions of a service provider risk interfering with this mission.

—The authorization of declaration procedures adopted for these services should be published not later than December 31 1992; before their implementation, the Commission will ensure the compatibility of these procedures with the provisions of the Treaty.

—Independently of the measures concerning packet or circuit switching of data, those member States retaining exclusive or special rights for the establishment and operation of networks must take the steps needed to publicize the conditions of access to the public networks in an objective, and non-discriminatory way. They were to communicate the scope of these measures to the Commission before December 31 1990. At the time of any increase in tariffs applicable to leased circuits, they must provide the Commission with justifying information. They were asked to ensure that the characteristics of the technical interfaces necessary for the utilisation of public networks were published by December 31 1990.

—The Directive also stipulated that, from July 1 1991, member States would arrange that the allocation of authorizations, the control of mandatory agreements and specifications, as well as the allocation of frequencies and the monitoring of their conditions of use, should be carried out by an entity that is independent of the telecommunications organizations. This is an application of the principle of separation of the regulatory and operational functions.

—Finally, the Directive stipulated that, for a period of three years, the member States should communicate information to the Commission in order to produce annual reports on the compliance of the member States. These reports will be communicated to the member States, the Council, the European Parliament, and to the Economic and Social Committee.

The Council Directive relating to the implementation of the provision of an open telecommuncations network (Open Network Provision, ONP) concerns the harmonization of the conditions of access to telecommunications networks, and, where appropriate, to public telecommunications services.

The Directive ensures that the conditions of network provision conform to a number of basic principles. These conditions are to be founded on objective criteria: they should be perfectly clear, should guarantee equality of access, and should not be discriminatory. They should not restrict access to the public telephone network or telephone services, other than for reasons based on four essential requirements:

—security of network operation,
—maintenance of the integrity of the network,
—interoperability of services, in the cases where this is justified,
—protection of data in appropriate cases.

These conditions will be defined in stages. They concern selected topics which are presented in an appendix to the Directive, including leased lines, packet and circuit switched data services, ISDN, voice telephony, telex, and in specific cases, mobile services. Two other topics have been added which require complementary studies:

—new types of access to the network, such as access to circuits connecting subscribers to the public network exchange for data services (Data over Voice) and access to new intelligent network functions.
—access to broadband networks.

Another appendix of the Directive sets up a time-table setting out its implementation over a three year period.

The current state of implementation in January 1992 is the following:

—Application of ONP to leased lines
In February 1991 the European Commission adopted a proposal for a Council Directive on the application of open network provision to leased lines. The first reading in the Council of Ministers and European Parliament has taken place and a common position has been adopted in the Council of Ministers of December 18/19 1991. The second reading is now on course in the European Parliament, after which the Directive can be formally adopted by the Council during the first half of 1992 and can come into force 12 months after its adoption.

Two major aspects of the provision of leased lines are the main focus of the proposal:

● the Community-wide harmonization of the usage conditions and tariff principles for all leased lines which are provided by Telecommunications Organizations
● the mandatory provision of a limited set of leased lines with harmonized technical standards across the community.

A minimum set of leased lines is listed and must be made available throughout the Community. The five types of leased lines identified represent the large majority of leased lines which are currently in use and in demand. A mechanism is provided to enable updating the list of leased lines on the basis of changes in market demand and in technology.

On the basis of a specific ONP mandate, ETSI is currently producing standards for the five types of leased lines proposed, which are due to be finalized and approved by the end of 1993.

Line type	Interface	Performance
Analog		
ordinary quality	2 wire or 4 wire	M1040
special quality	2 wire or 4 wire	M1020/M1025
Digital		
64 kbit/s	G703	Relevant G800
2 Mbit/s unstructured	G703	Relevant G800 Rec
2 Mbit/s structured	G703/G704	Relevant G800 Rec + in service monitoring

Only limited restrictions on the use of leased lines will be allowed based on the respect of exclusive or special rights and on compliance with essential requirements. Guidance on the Community wide interpretation of the essential requirements is provided, thus harmonizing the respective national regulations in this respect.

The proposal also sets out the requirement of users to be able to order leased lines in a common fashion, and where requested, to be able to communicate with a single Telecommunications Organization for ordering and being billed for intra-Community leased lines.

The basic principles of cost orientation and transparency are to be applied to leased line tariffs. Telecommunications organizations are obliged to use cost accounting systems which are suitable for the verification of these principles.

In addition the principle of subsidiarity is embodied in the proposal by emphasizing the important role of the national regulatory authorities of the Member States in the implementation of this directive.

—Application of ONP to voice telephony
An Analysis Report on the application of Open Network Provision to Voice Telephony was issued for public comment in July 1991. This Report presented the main proposals in a concise form in order to facilitate the consultation process. The Comment period closed at the end of October 1991. A draft proposal for a council directive was then prepared based on the text of the Analysis Report including a large number of amendments suggested in the comments received.

The draft proposal, currently under discussion at the ONP Committee, embeds three basic goals for the application of ONP to voice telephony:

• to establish the right of users when dealing with Telecommunications Organizations

- to provide open and non-discriminatory access to the telephone network infrastructure for competitive service providers and other telecommunications operators (e.g. mobile)
- to support the demands of the single market, particularly in the provision of European-wide telephony services, and in the planning and coordination of pan-European numbering.

The scope of ONP for the voice telephony area extends to cover both the service and the network over which the service is provided, covering the right of connection to and use of the public network regardless of the underlying technology.

National regulatory authorities are called to ensure that:

- information on services offered, conditions for attachment, usage restrictions, target dates for new services and numbering is regularly published and up-dated
- targets are set and published for service's supply time and a number of quality of service indicators
- the performance of telecommunications organizations in relation to those targets is published on an annual basis
- users have contracts specifying service quality levels and rights to compensation
- targets are set and published for the provision of itemized billing allowing users to check their bills.

The draft proposal also covers the more advanced requirements of service providers and other operators, including the important issues of interconnection and special network accesses.

A minimum set of voice telephony features and services harmonized across EC-Member States, according to relevant ETSI standards, is a key stone of the draft proposal.

MINIMUM SET OF FEATURES AND SERVICES

- Call transfer (of an established call)
- Call forwarding (if busy, no reply, unconditionally)
- DTMF operation
- Direct dialling in (multiline arrangements)
- European-wide kiosk billing arrangements
- European-wide access to freephone numbers
- Automatic reverse charging
- Calling line identity
- Cross-border access to operator services
- Cross-border access to directory enquiry services

From a harmonization perspective the draft proposal calls for common technical specifications for the telephone network interface, including the socket, taking into account the evolution of network technology (PSTN to ISDN).

Tariffs should take into account economies of scale associated with bulk provision and network load, e.g. off peak rates. Special tariffs may be agreed with the national regulatory authority for provision of socially desirable services such as emergency services, or for low usage users or specific groups like handicapped people.

The Commission also proposes that national regulatory authorities are made responsible for the control of national numbering plans. Provisions are also made to coordinate the future evolution of national numbering plans at a Community level.

Interconnection, special network accesses and cross-border interworking issues are recommended to be established on basis of commercial and technical agreement between parties with intervention by the regulator when and where required.

The Commission's proposal on the application of ONP to Voice Telephony is planned to be submitted for approval to the Council in the first quarter of 1992.

—Application of ONP to packet switched data services (PSDS)
In June 1991 the European Commission adopted a proposal for a Council Recommendation on the harmonized provision of a minimum set of Packet Switched Data Services in accordance with Open Network Provision principles. A timetable is set for the provision of these services, where they are not yet available. Quality of service indicators are also proposed.

For the harmonized minimum set, reference is made to the list of packet switched public data networks standards suitable for ONP published in the Official Journal on December 29 1990. A mechanism is provided to update this set on the basis of changes in market demand and in technology.

Access type	Service
Direct access	X25
Indirect access	X28
Indirect access	X32

The proposal expresses the requirement of users to be able to order Packet Switched Data Services in a common fashion, and where requested, to be able to communicate with a single organization for ordering, billing and maintenance purposes.

The supply conditions for Packet Switched Data Services are to include at least a number of parameters which are of vital importance to users, e.g. the delivery period for a type of Packet Switched Data Service, the duration of the contractual period and the repair time. It also refers to the refund policy and the network performance targets.

Quality of service of the Packet Switched Data Service is addressed and the adoption of common indicators for the network performance aspects of quality of service, and corresponding measurements methods, are called for.

It is anticipated that the Recommendation will be adopted in the first half of 1992 during the Council's Portuguese Presidency.

—Application of ONP to Integrated Services Digital Network (ISDN)

A draft proposal for a Council Recommendation on the provision of harmonized ISDN access arrangements and a minimum set of ISDN functionalities in accordance with open network provision principles was discussed with the ONP Committee in September 1991, and was adopted by the Commission on December 13 1991.

The Commission has already published in the Official Journal on December 29 1990 the initial list of ISDN standards suitable for ONP.

The harmonized access arrangements concern the interfaces at the CCITT defined reference points, i.e. the S/T Interface point. Future access arrangements will be studied by ETSI and the CEC and the situation reviewed at the latest by December 31 1992. Under consideration in this respect are studies on the application of ONP principles to the M- and U-type interfaces.

Minimum set of access arrangements and ISDN functionalities available by January 1 1994	
Access	• Basic rate (2B + D at S/T ref point) • Primary rate (30B + D at S/T ref point)
Bearer service	• 64 kbit/s unrestricted (circuit mode) • 3.1 kHz audio (circuit mode)
Tele services	• Telephony 3.1 kHz • Calling line identification presentation
Supplementary service	• Calling line identification restriction • Direct dial in • Multiple subscriber number • Terminal portability

The access arrangements and ISDN services above must be available by 1994 in every EC-Member State according to the relevant ETSI standards published in the Official Journal. Additionally, regarding the extended set below, Members States should encourage telecommunications organizations to publish target dates for availability.

Extended set of ISDN functionalities	
Bearer services	• 2 × 64 kbit/s unrestricted (circuit mode) • 64 kbit/s unrestricted (reserve/permanent mode) • Packet bearer services over the B and/or D channels (packet mode)
Supplementary services	• Call transfer • Call forwarding • Reverse charging • Freephone • Kiosk billing • Closed user group • User to user signalling • Malicious call identification • Network management services

Common indicators for the network performance aspects of the quality of service of ISDN bearer services and common measurement methods are to be adopted.

This Recommendation will be subject of Council discussion during the first half of 1992 and it is anticipated that it may also be adopted during the Portuguese Presidency.

—Conclusion

The further definition of ONP in concrete measures is a staged process, based on studies, analysis, consultation of interest groups and consensus at the political level. It is a dynamic process which takes into account the continuous changes in technology and in markets. In this respect studies will follow the new developments in Intelligent Networks, Broadband Communications including Asynchronous Transfer Mode (ATM) switching techniques (in cooperation with RACE), Mobile Communications, Metropolitan area networks (e.g. FDDI—Fiber Distributed Data Interface, DQDB—Distributed Queue

Dual Bus) and advanced transmission techniques (e.g. SDH—Synchronous Digital Hierarchy).

● *At the national level, a steady flow of legislation was generated*
While the introduction of competition into the telecommunications markets of the Member States began before the publication of the Green Paper in 1987, the process of adjusting national legislation in accordance with market reform has recently entered into a new phase. Following the adoption of Community Directives implementing the principles of the Green Paper, Member States have substantially modified their national legislation, or are in the process of doing so, to make it compatible with Community law. The following is a description of the changes that have taken place and those currently underway in each of the Member States.

Common fundamental principles can be found in each of these laws:

—the separation of the regulatory functions from the networks operation;
—the liberalization of terminal equipment;
—the liberalization of services, and
—the maintenance of special rights over the network infrastructure and telephone services.

However the methods used to apply these principles vary.

1. Belgium

On March 21 1991 the Belgian parliament adopted the *Loi portant réforme de certaines entreprises publiques économiques* which split *Régie des Télegraphes et Téléphones* (RTT) into two separate organizations for the regulation and operation of Belgian telecommunications. Belgacom is to become the sole Belgian public network operator and the *Institut belge des services postaux et des télécommunications* (IBPT) will provided regulatory oversight. Belgacom remains a State-owned company controlled by the Minister for Post and Telecommunications.

The new law also makes a division between reserved and non-reserved services. Reserved services are to be provided exclusively by Belgacom which also maintains a monopoly over the operation and maintenance of the network infrastructure. These reserved services include: telephony, telex, mobile telephony and paging. Moreover, Belgacom is the exclusive operator of the only public switched data network in the country. This exclusive concession is due to expire on December 31 1992.

2. Denmark

The organizational structure of Danish telecommunications was substantially altered in 1987 when a number of tasks were transferred from the PTT to local operating companies, primarily for inland services. At the time, Telecom Denmark provided the trunk network between these operating company areas and all international services for the country. In addition, the regulatory and operational functions of the PTT were separated.

On November 14 1990 Act no. 743 introduced further regulatory changes. The existing five companies were consolidated into a new holding company in which the government currently maintains a 51% stake, with 49% held by private investors. Telecom Denmark was transformed into a limited liability company, Tele Danmark A/S, which has been granted an exclusive licence for the installation and operation of all telecommunications services for transmission routes and exchanges linked to telephone services, text and data communications services, and conveyance through the telecommunications network of radio and television programs. Act no. 744, issued at the same time (November 14 1990), created one exception to this exclusive licence by empowering the Minister of Communications with the authority to grant licences for two competing GSM mobile network operators. One of these licences will be granted to Tele Danmark A/S.

Competition in mobile services is to begin March 1 1992. The licence will last up to 5 years and can be revoked with a one year's notice. VANS used in the provision on GSM will not be subject to any regulation and the provision of value added services in general has been liberalized in Denmark since July 1 1988.

3. France

On May 19 1989, the *Direction de la Réglementation Générale* (DRG) was enacted by Decree, charged primarily with the responsibility of granting licences. On July 2 1990 the post and telecommunications responsibilities were separated within the Ministry by law No. 90/568. Then on December 29 1990, law No. 90/1170, which amended the French *Code des Postes et Télécommunications*, brought French law into line with Community Directives by clearly separating the PTT's regulatory and operational functions. France Télécom was transformed into an autonomous operator, no longer receiving financial contributions from the State to its general budget. Under this new regulatory regime, France Télécom has retained its monopoly over the establishment and operation of public switched networks. Furthermore, France Télécom has an exclusive monopoly over the provision of telephone, telex and data services. Basic data services, however, are to be liberalized by January 1 1993. Value added services provided on leased lines are open to competition and require either a simple declaration or notification for licensing. Satellite and mobile services are also open to competition and licences are granted on a case-by-case basis.

4. Germany

The separation of the regulatory and entrepreneurial functions of telecommunications in Germany was brought about by the Postal Constitution Act (*Postverfassungsgesetz*) of June 8 1989. Operational tasks are currently undertaken by the public enterprise, Deutsche Bundespost Telekom (DBP Telekom), and regulatory oversight is provided by the Federal Minister of Posts and Telecommunications. On July 1 1989 the Telecommunications Installations Act (*Fernmeldeanlagengesetz*) was amended by the Act on the Restructuring of the Postal and Telecommunications System (*Poststrukturgesetz*) which has brought about the partial liberalization of the telecommunications market through the elimination of the formerly sweeping rights of the German Federal Post Office, Deutsche Bundespost. Despite the fact that a monopoly concession still exists for the transmission paths of the fixed terrestrial public network and voice telephony services transmitted through this network, the reform has introduced the principle that monopoly rights are to be considered an exception and, as such, must be justified.

In an effort to justify and clarify DBP Telekom's monopoly, a number of ordinances and administrative rules have been adopted. On November 27 1990, the Minister of Post and Telecommunications issued an administrative rule by which the exercise of the Federal Government's network monopoly was officially transferred to DBP Telekom. On the basis of the amended Telecommunications Installations Act, the Federal Government adopted the Telecommunications Ordinance, *Telekommunikations-verordnung*, "TKV", of June 24 1991, which sets out the legal framework for the provision of "monopoly services" by DBP Telekom along with basic rules for the provision of competitive services by DBP Telekom. This ordinance also contains provisions which are intended to serve as the legal basis for the implementation of the European Community's Open Network Provision rules. Most recently, the Federal Ministry has adopted the "Administrative Rule for the definition of the power to exercise the network monopoly as transferred to Deutsch Bundespost Telekom", of September 19 1991. Administrative rules for the definition of the voice telephone monopoly are presently being prepared by the Ministry. The one notable exception to the DBP Telekom monopoly is the provision of voice services via satellite to the territory of the former GDR.

Aside from DBP Telekom's monopoly on telephony, all other telecommunications services can be freely provided over fixed or switched connections to be made available by DBP Telekom. In particular, the provision of mobile (radio) networks has been opened to competition with the licensing of Mannesmann Mobilefunk on February 15 1990. DBP Telekom was also awarded a mobile licence by Administrative Rule at the same time. Several licences have also been granted for local mobile networks (Bündelfunk). These operators are entitled to interconnect their base stations and network management installations for a region with DBP Telekom's voice telephony service, through transmission paths established by DBP Telekom. Six such licences have been granted so far.

Packet and circuit switched data services, and value added services, are open to competition. Satellites are regulated under a dual regime, one intended for the transmission of data and the other for the rest of satellite services. Under the former, a satellite service provider will be licensed if the service "does not affect radio communications", and the latter will only be denied if it serves to by-pass voice communications on the terrestrial network. At present, 18 satellite licences have been awarded.

5. Greece
The network operation and the provision of telecommunications services remain the exclusive right of the State-owned telecommunications organization, OTE S.A., and is regulated by the Ministry of National Economics and the Ministry of Transport and Communications. Liberalization of VANS is currently being discussed and the Ministry of Transport intends to privatize 49% of OTE. Greece is the only Member State that does not maintain a mobile telephony system.

6. Ireland
According to the Postal and Telecommunications Services Act of 1983, Telecom Eireann, a statutory company wholly owned by the Irish government, has the exclusive right to provide domestic telecommunications services. The Act also permits Telecom Eireann to grant licences to individual telecommunications service providers, subject to the Minister of Communications' approval. This would be based

on public interest requirements, consistency with Telecom Eireann's provision of the national service, consistency with population and technological growth, and an emphasis on the need for cross-subsidization. One such licence has been awarded to Nexus Communications for the provision of Audio Conference Services.

The Minister of Communications is also permitted to grant certain licences following consultation with Telecom Eireann and the Minister of Finance. Under this provision of the 1983 Act, "An Post" (the Irish postal company) was issued with two licences to provide electronic mail services nationally and internationally. A licence of this type was also awarded to Telecom Eireann for the provision of all international telecommunications services to and from Ireland.

Ireland has no competition for value added services and Telecom Eireann retains a controlling interest in the county's cable TV networks. Proposals are currently being considered to amend Irish legislation, bringing it into compliance with Community law. This new legislation would reinforce Telecom Eireann's monopoly, but would also provide procedures for licensing value added services and third party resale of lease lines which will be allowed from the end of 1992. The legislation will also make provisions for ensuring the separation of regulatory and operational functions.

7. Italy

The general framework for telecommunications legislation is provided by the *Decreto del Presidente della Repubblica 29-03-1973, No. 156*, commonly referred to as the "Telecommunications Code". It has recently been updated by the *Decreto Ministeriale 6-04-1990, Approvazione del Piano Regulatore Nazionale delle Telecomunicazioni* (Telecommunications Reference Plan Approval) of April 6 1990, which provides technical definitions for network architecture and function, and classifies services.

The 1973 Decree establishes that telecommunications services are an exclusive State property, and that the PTT Ministry is empowered with the provision of these services. The Decree gives the Ministry the choice of providing the telecommunications services itself or by means of licences (*concessioni*). Technically a call for tenders must be made in the process of finding qualified public network operators, but this procedure may be avoided, and a licence directly granted, when the concessionaire is a company in which the State is the major shareholder. This latter option has been, and remains, the chosen approach to establishing the public network in Italy.

There are four State-controlled entities which maintain the exclusive rights for the installation and exploitation of the public telecommunications network and service: Sip (domestic service), Italcable (intercontinental), Telespazio (satellite) and *Azienda di Stato per i Servizi Telefonici* (ASST) which is a fully State-owned company that controls some of the domestic long-distance and intra-European network, but is likely to disappear as its competences are absorbed by Sip. These exclusive concessions are limited to 20 years and undergo periodic re-examination, the next being scheduled for 1992. A second licence was also issued to Sip in 1989 (*Convenzione aggiuntiva il Ministero PT e la Sip*) that grants a monopoly concession over the implementation and exploitation of the switched packet data network.

While there is a desire to integrate these separate entities and split the operational and regulatory function of the State, this has not yet taken place. Furthermore, neither the public licences, nor the global legislative framework, refer (as yet) to global transparency or non-discriminatory requirements. Both of these issues have been under discussion. In fact, two bills addressing the issue of regulatory and operational separation have been under discussion since 1988, but have not been approved. A bill addressing transparency and equal access requirements was recently abandoned after it was found not to be in line with EC directives issued in the meantime.

8. Luxembourg

On September 29 1990 Luxembourg brought its telecommunications law into line with Community Directives through the adoption of the *Règiement grand-ducal du 3 août 1990 fixant les dispositions générales applicables aux services publics de télécommunication.* This new law defines the boundaries between monopoly services and competitive services, opens value added services to competition and establishes an independent approvals body. The law reinforces the PTT's monopoly rights over telephony, telex, teletex, packet and circuit switched data transmission, mobile telephony and radio telephony. The PTT has also been transformed from a State institution into a public firm, and value added services can be provided without a licence or declaration.

9. Netherlands

The Dutch PTT (PTT Nederland N.V.) has been granted an exclusive concession for the construction, maintenance and exploitation of the telecommunications infrastructure based on the Dutch Telecommunications Act (*Wet op de Telecommunicatie Voorzieningen*, "WTV") which came into force on January 1 1989. Under the new law, the regulatory and entrepreneurial functions of the PTT have been divided between the Ministry of Transport and Public Works (MTPW) and PTT Nederland N.V. respectively. PTT Nederland was also transformed into a public limited liability company, with the State currently holding all existing shares.

The Act requires the PTT to provide certain mandatory services relating to the "direct transport of data". These PTT-mandatory services include mobile telephony, data transport, mobile data transport, telegraph and telex. Leased lines must also be considered as part of the mandatory services, since the PTT is required to provide them to anyone with equal consideration, subject to certain conditions. The Act also requires the public utility and private commercial activities of the PTT to be separated in order to prevent cross-subsidization of reserved services and those open to competition. This process must be completed by January 1 1994, but may be extended by 2 years by Decree.

Competition in value added services is currently allowed; however, VAS providers who use leased lines or private networks receive licences only if the services do not compete with the mandatory services assigned to the PTT. In practice, because the Minister of Transport takes the view that the PTT is not subject to the same licensing requirements of value added service providers, the PTT has the opportunity to offer so-called "combined" services (e.g. mandatory services combined with value added services), providing a package in the market place that no other service provider can compete with. Thus, value added services may be opened to competition, but little actually takes place.

On January 21 1991, the Minister of Communications informed the Dutch Parliament of his intention to introduce a second (maybe third) operator for mobile (telephone) communications. Furthermore, the Minister has also supported the Commission's proposal for liberalizing the earth segment of satellite communications.

10. Portugal

Law No. 88/89, the Basic Law on Telecommunications of September 11 1989, establishes the State-owned telecommunications network to provide basic telecommunications services under a monopoly concession. Basic services include fixed telephony, telex and data transmission services.

Three operators share in the total provision of the network infrastructure and basic services. *Correios e Telecomunicacoes de Portugal* (CTT), a state owned monopoly, provides telephone and telex services on a nationwide basis except in Lisbon and Oporto; *Telefones de Lisboa e Porto* (TLP), a publicly owned company (soon to be partially privatized), has exclusive rights over the provision of infrastructure within Lisbon and Oporto; and *Companhia Portuguesa Radio Marconi* (CPRM), owned primarily by the State for the provision of international services (except in Europe) and has *de jure* exclusivity on the installation and operation of all satellite services (including VSAT). In 1985, CTT and TLP developed Telepac, designed to provide packet data services nationwide.

Non-basic services are consider liberalized and classified as "complementary services" and "value added services". Complementary services are those associated with "complementary networks" or networks based on leased lines, but with their own switching or processing unit. These would include mobile GSM, radio paging, and videotex. Value added services refers to telefax, voice mail and audiotex services. Access and usage conditions for value added and complementary services are established in the Decree-Laws No. 329/90 of October 23 1990 and 346/90 of November 3 1990, respectively. Value added services are subject to a simple authorization procedure, and complementary services using fixed networks must meet certain technical, economic and financial obligation to receive a licence. Four value added service providers have been authorized since March 1991.

The process of separating regulatory and operational functions has been aided by the re-activation of the *Instituto das Communicacies de Portugal* (ICP) on August 23 1989 by Decree-Law No. 283/89. The institute, originally established in 1983, is designed to serve in an advisory capacity to the Ministry of Public Works, Transport and Communications, and has been given the responsibility of overseeing public operator obligations, as well as establishing licensing and registration for new fixed complementary and value added services, and the tender for GSM operators.

Law No. 88/89 limits the participation of foreign companies in the capital of public or complementary telecommunications carriers to 25%.

11. Spain

The basic legal framework for telecommunications in Spain is provided by the Telecommunications Law 31/1987 (*Ley de Ordenación de las Telecommunicaciones*, "LOT") of December 18 1987. LOT not only distinguishes between basic and value added services, but also provides for the separation of the regulatory and operational functions of the telecommunications monopoly. The regulatory authority is provided by the *Direccion General de Telecommunicaciones* (DGT), of the Ministry of Works and Transports, which is responsible for the provision of telegraph, telex and basic telephone services. The DGT has granted the exclusive right of providing these services to *Telefónica de Espana* (a 47% State-owned company). Foreign participation in *Telefónica* is limited to a maximum of 25% of total investment.

The DGT is now required to offer carrier circuits, switched networks and leased lines to value added service providers, granted by administrative "authorization" or "concession". No general regulation for value added services has been issued, but the Decree on the use of the Radioelectric Domain, July 7 1989, which concerns technical and frequency requirements for radio transmission, currently applies. Concessions for value added services will only be granted if the service cannot be provided by the public network. This concession can be revoked if at any time it is determined that the public network could effectively substitute for the special installation. Additional regulations for value added services, that have not yet been published, will establish a ratio for each service of the total cost of the support service used, to the gross revenues in offering the service. This is designed to ensure that value is being added to the underlying transport. Transparency for value added services will be provided through publication of the Central Register maintained by the DGT.

12. United Kingdom
The Telecommunications Act of 1984 ended the network monopoly of British Telecom, and began the withdrawal of State intervention to a degree unprecedented in Europe. The State's role is now limited to facilitating competition, outlining technical standards and ensuring the provision of high quality universal service. Presently, two licences are granted for the operation of nation-wide fixed terrestrial public networks, held by British Telecom and Mercury Communications Limited. A third licence is held by Kingston Communications and is limited to the provision of services to the city of Kingston upon Hull. The licensing of further network operators was called for by the White Paper, *Competition and Choice: Telecommunications Policy for the 1990s* of March 25 1991, commonly referred to as the 'duopoly review'. Not only was the White Paper designed to modify the licences of BT and Mercury, but it contained proposals to increase competition in all segments of the market for the provision of telecommunications services. This was reflected by its support of the principle of equal access and by allowing mobile network operators other than BT and Mercury to offer fixed services.

The regulatory authority in the United Kingdom is provided by the Secretary of State and the Director General of Telecommunications who is the head of OFTEL and has the power to make references to the Monopoly and Mergers Commission with respect to modifications of licences. Since the Act provides no procedural qualifying criteria for a licence, nor criteria as to how competing requests are to be resolved, the Director of OFTEL is left with wide discretionary powers.

Most value added and competitive services are permitted in the United Kingdom and do not require any further registration under the Branch System General Licence (BSGL) scheme. The BSGL was first introduced in August 1984 and was since revised on November 8 1989, expanding the range of services which it covers. Approximately 850 services, including satellite and mobile services, exist under the BSGL. The United Kingdom has a wide range of local cable networks, numerous mobile network operators, and in 1989 granted licences for the operation of three new Personal Communications Networks.

Special individual licences are required for services on cable networks, via one-way satellites, and via two-way satellites. In the case of reserved services, fixed voice, mobile voice, or intelligent networks, applicants have generally written to OFTEL with an explanation of their proposal. If OFTEL feels it necessary, the service providers are issued a special licence; otherwise they will be covered under the BSGL.

It should be noted that the BSGL does not permit international simple resale service, nor interconnection to an international private circuit which is to carry messages that have been or are to be conveyed by a telecommunications system run by a third party operator in the UK or by a public switched network. These conditions are to be modified following the duopoly review.

All licences (including the BSGL) are limited to a 25 year duration except the Kingston licence which expires June 24 2009.

3.4.4.2 *Regional cohesion and the social dimension*

The formation of a balanced Europe requires the development of its disadvantaged regions, which in turn requires the establishment of advanced infrastructures—particularly those for telecommunications. This is the objective of the STAR program (Special Telecommunications Action for Regional Development).

The Council adopted this program on October 27 1986; it is supported for a period of five years by a grant of 1.5 billion ECUs, of which 780 million comes from the Community budget, while the remainder is provided by the beneficiary States.

The program supports the financing of the infrastructures necessary 'for the establishment and promotion of advanced telecommunications services in Greece, Southern Italy, Ulster, the Republic of Ireland, Spain, Portugal, the French Overseas Departments and Corsica.

Investments relating to advanced telecommunications infrastructure represent 80% of the total budget. They concern the main interregional and international transmission arteries (employing optical fibres or satellite transmission), the digitization of networks with a view to the accelerated introduction of ISDN, data transmission networks, and mobile communication networks. Control and certification laboratories have also been created.

The remaining 20% has been invested in applications: feasibility studies, promotional activities and demonstrations, operational applications in varied sectors (tourism, commerce, distribution, etc.).

The implementation of these investments calls on technical operators (network operators or telecommunications engineering companies), or economic operators (marketing study companies, local mixed economy companies). The promotional operations are aimed primarily at the development of small and medium local businesses and industries.

Following the approval of an investment program by the European Communities Commission, activities are managed by the beneficiary member State, who must put in place an appropriate organization. Depending on the State, the establishment of such an organization has taken from 6 to 18 months.

To provide an overview of the achievements of the STAR program we cite the development of large capacity interregional and international links (in Ulster, Greece, Italy, the French overseas Departments, and Corsica), the digitalization of telephone networks (in all the countries in the program), and pilot ISDN projects for user access (in Corsica, the Republic of Ireland, and Spain).

STAR also supports the establishment of cellular radio-telephone networks (in Corsica, the Irish Republic, Portugal and Greece), data communications (by packet switching and Videotex in all the countries involved, X-400 messaging in Ireland and pre-ISDN in the Irish Republic, Italy and Ulster). Laboratories for control and equipment certification were set up in Spain and in Italy.

Infrastructure operations have been integrated into the investment programs of national telecommunications operators, and for the most part have rapidly become operational. Even if STAR finances only a small percentage of the overall investment of the operators, its effect is nevertheless very significant at the regional level. In addition, it has often led the network operators to earmark their investments in such regions.

To bring about the transition between the current and the subsequent phases of STAR (the subject of a proposal put to the Council in 1991), on July 25 1990 the Commission launched the "Telematic" program, which was provided with a budget of 200 million ECUs over a period of two years. This program focused on the implementation of applications, thereby augmenting STAR, which had only allocated 20% of its resources to this field.

Because STAR is managed within the framework of FEDER (European Funds for Regional Development), the beneficiary member State must provide 45% of investments and organize the implementation of the actions. The program sets up "joint management" systems between parties with diverse natures and interests. These include regional, national and Community powers; economic operators; planning organizations having social responsibility, etc. It should be added that the program is subject to political and economic difficulties existing in the countries where it is applied. Therefore, time (an average of one year) is needed in order to establish a viable organization, to define projects, and start work. Despite these difficulties, STAR has acted as a powerful catalyst for the offer and demand for services in those regions where it has been implemented.

3.4.4.3   Taking into account the social dimension

In the context of the technical upheavals and changes in organizational structures characterising telecommunications today, it is essential not to lose sight of the impact of these many rapid evolutions on the social level. In the first place, they affect employment with regard to both the number of positions available and the qualifications required to fill them; they are also in the process of profoundly changing ways of life and social attitudes.

In order to obtain a consensus on these changes and on the creation of a "European telecommunications zone", it is necessary to study their influence on various social factors in order to predict potential short term harmful effects of these changes, and to remedy them to the extent possible. In this context, a dialog with the social partners involved takes on a fundamental importance.

Since the beginning of the process of defining and establishing a Community telecommunications policy, frequent meetings and contacts between the Commission and union representatives for the sector have been organized. Orientations on the subject of liberating services have been held, and plans for measures to be proposed to the Council have been discussed. In cooperation with the unions, a study has been completed concerning changes in telecommunications employment. This study predicted a slight increase in industry-wide employment (from 1.8 million in 1987 to 1.9 million in 1992), with a growing proportion of jobs in the services area (71% in 1980 compared to 76% in 1992). The study also concluded that restructuring and greater flexibility could be anticipated in the employment field; that a higher level of qualifications will be required; and that there will be an increasing need for professional training.

In October 1990, in order to more effectively deal with the problem of social adaptation to the industry's evolution, the Commission set up a joint committee composed of union representatives, employers and the Commission. A dialogue had thus been institutionalized, and the effectiveness of Community action will be strengthened by the resulting increase in opportunities for coordination.

3.4.4.4   *External aspects of the telecommunications policy*

In several instances, we have already seen that practically all aspects of the development of Community telecommunications have been influenced by the same events affecting the rest of the world. Conversely, Community policy influences the evolution of telecommunications in other countries. Telecommunications is therefore experiencing a growing "globalization". The world economy is fueled by the information that it now carries; the world market for telecommunications equipment and services is considerable (already more than 500 billion ECUs per year), is growing continuously, and is subjected to increasingly active competition. Technological supremacy is fiercely contested among all industrialised countries, and the stimulating economic impact of telecommunications renders it indispensable to the progress of developing countries.

It is therefore not surprising that the external political aspects of telecommunications are assuming an ever-increasing importance. It is appropriate for the Community—vested with direct authority, particularly in the matters of international commerce, standardization policy, and cooperation with developing countries—to be active in this domain. The Commission acts on behalf of the Community in these areas. In the multilateral and bilateral relationships that it maintains with a large number of countries and organizations, it is supposed to take into account all the complex aspects of Community interest in this sector. It strives to promote an open international business environment, but given the high stakes, it should also pay close attention to fixing the future rules of the game. The recent political evolution of central and eastern Europe has provided a new dimension to this problem.

● *Multilateral considerations*

—GATT: Telecommunications in the Uruguay Round   Since its creation, the General Agreement on Tariffs and Trade (GATT) has been the forum for discussion of agreements concerning international commerce. The Community's interests in GATT are defended by the Commission in accordance with the mandates conferred on it by the Council.

In September 1986, a new series of negotiations was launched at Punta del Este, in Uruguay, inaugurating what was called the "Uruguay Round". Telecommunications was included in four areas of these negotiations: services, public markets, technical obstacles to trade (standards), and questions of access to markets (customs tariffs and non-tariff barriers). Agreements were sought by the end of 1990. (It is still not settled. The negotiations are continuing; the Americans are not yet ready.)

The service domain was a new category for GATT, which had until then only dealt with trade in goods. The discussions first centred on defining a frame of reference, arising from the General Agreement on Tariffs and Trade, that could be applied to all trade and services. They then took particular sector-dependent aspects into account. The discussion of the applicability of these principles to telecommunications began in June 1989, and has been pursued with modest progress. They deal with the nature of services to be included within the scope of trading practices (the distinction between basic services and value added services), and with divergences of regulatory structures. There exists a sufficient global core of convergent points (transparency, conditions of access, and network use), for an agreement to be drawn up. Nevertheless, important difficulties remain (the issue of inter-operability, protection of personal data, and access to information), and the developing countries are preoccupied to the extent that they need time to develop their industries and their domestic markets before allowing foreign companies to penetrate their markets.

With regard to access to the equipment markets of public telecommunications network operators, the negotiations center on the widening of the GATT "public markets" code, and eliminating certain non-tariff barriers. In September 1990, the Community presented a list of requests concerning non-tariff barriers to the United States and Canada, among others. The objective of the market code discussions is an expansion to cover other fields, including telecommunications, which do not currently figure in its scope (with the exception of NTT, for a very restricted range of products).

In August 1990, the Community formally introduced its negotiating position; a request that all public or private organizations operating public telecommunications networks under special or exclusive rights be subject to discipline similar to that imposed in the EC by the Directive on the opening of markets to network operators, thus ensuring transparency and non-discrimination in the awarding of public works contracts.

It was intended that this arrangement would apply to contracts from AT&T and the North American BOCs on the same basis as for those from European companies, as well as to all the contracts of NTT including strategic network products currently purchased by NTT outside the GATT procedures. No agreement had yet been reached on these issues!

In the area of technical obstacles to trade, the Community strove to establish a better balance within the framework of the GATT agreement by extending its scope, to include the adoption of international standards and to ensure the transparency of specifications adopted by local and private standardization organizations.

Finally, with regard to customs tariffs, a "halfway review" of issues that had appeared since the start of the "Round" was held in December 1988 in Montreal. A number of principles and objectives were established on that occasion, and implementation was started in July 1989.

—Relationships with the International Telecommunications Union (ITU)
In 1989, the Community obtained observer status at the ITU. The Commission participated actively in the meetings of the International Radiocommunications Consultative Committee (CCIR) during discussions on the world standard for high definition television, contributing to the development of the efforts of European industry in this area.

The Community also took part in the World Administrative Telephones and Telegraphs Conference (WATTC) (1) in November 1989, where a new round of international regulation was negotiated. Due to the coordination between the Commission and the member States, a joint declaration was submitted by the States indicating that they would abide by this regulation in keeping with their obligations under the terms of the Treaty.

—Relationship with the Organization for Economic Cooperation and Development (OECD) The Commission takes part in discussions on telecommunications, which are dealt with by the OECD's Information, Computing and Communications Policy Committee (ICCP).

---

(1) or CCIR.

This committee's attention has essentially been centered on trade in telecommunications services, in preparation for the Uruguay Round and its discussions of individual sectors.

● *Bilateral relationships*

—*Relationships with the United States*   Relationships with the United States have been discussed in detail in previous chapters. It will be recalled here that they concern arrangements on telecommunications matters within the terms of the Trade Act, access to the American and Community markets, and respective approaches to questions of standardization, public markets, and terminal and services markets. Discussions with the United States on these questions have been ongoing since 1986. As has already been indicated, the designation of the Community for implementing the "Omnibus Trade Act" has led to several meetings where the parties involved have been able to proceed to an exchange of information on problematic issues. The Commission believes that negotiations in this area should be handled within the framework of the Uruguay Round, and also wishes to bring Community concerns about the difficulties of access by European industries into American markets into the debate.

—*Relationships with Japan*
The Commission has periodically met with the Japanese authorities since 1987, and an agreement has been reached to deal with telecommunications during high-level meetings on a nominally annual basis. The Commission also has regular contacts with NTT. These discussions center on exchanging regulatory technical information, encouraging Japan to apply international standards, and questions of reciprocal market access. Specific working groups have been created in the fields of networks, high-definition television, and mobile communications.

The Commission is preoccupied by the large and growing trade deficit between the Community and Japan. Efforts have been made by NTT (in several information seminars) to motivate European industry to submit bids to tender on the Japanese market, but the results have remained marginal. The Japanese market for network equipment, in effect, remains entirely dominated by the "NTT family", composed of NEC, Fujitsu, Hitachi, and Oki Electric, who share it according to well defined percentages.

European manufacturers, discouraged by past difficulties, do not consider the Japanese market as justifying a large effort because the results in general have been disappointing. This attitude poses a problem for the pursuit of discussions with Japan, who, for its part, appears to have adopted a policy towards openness which exists more in principle than in reality.

—Relationships with the EFTA countries
The European Community is establishing increasingly close cooperative links with the countries of the European Free Trade Association (EFTA) formed by Norway, Sweden, Finland, Iceland, Switzerland, and Austria. This spirit of cooperation had already been affirmed in the "Declaration of Luxembourg" adopted by the Ministers of the Community and of the AELE on April 9 1984, and consultations to clarify the terms have since taken place. However, the acceleration of Community evolution due to the introduction of the Single Act has also led to an accelerated evolution of links with the EFTA. On December 19 1989, the Ministers of the two groups decided to rapidly negotiate an agreement creating an "European Economic Zone" by January 1 1993, the same date as the internal Community market. This implied that an appropriate cooperative political framework would need to be finalized and ratified by that time, allowing the free movement of people, goods, capital and services, as established in the Community, to be extended to the EFTA. Such an agreement was successfully concluded in October 1991. The accompanying measures planned by the Community are to be mirrored by equivalent measures in the EFTA countries, as in the existing legislation which constitutes the "Community experience".

In the telecommunications sector, where significant cooperation with the EFTA has already begun (particularly in the areas of standardization and research and development), the "European telecommunications zone" can be expected to encompass the EFTA countries after 1993.

—Relationships with other countries
The Community maintains links or has relationships in the field of telecommunications with a number of other countries, primarily Canada and Australia, usually in the context of general cooperation in industrial and information technology matters.

—Relationships with developing countries
The question of telecommunications in developing countries is increasingly raised in multi-lateral forums, particularly at the ITU. In this context, a Bureau of Development of Telecommunications has been recently created. Relationships between the Community and developing countries mainly concern countries in the Lomé Convention

(ACP), Latin America, India, China, and Mediterranean countries.

For the ACP countries, the Community has financed a study concerning the use of satellites for rural communications, as well as improvements and extensions to the national and regional networks existing in Africa and the Pacific. However, for many African countries, the telcommunications sector represents a lower priority than the more traditional development fields.

Exchanges of information on Community policy and technology are in progress with the Latin American countries. Accordingly, numerous meetings have taken place since 1987 with experts in Brazil, Argentina, and Uruguay, particularly on the subject of narrow band ISDN. An awareness program concerning the most advanced techniques is under way with Mexico, and a training program for engineers is being drawn up.

With regard to India, several training courses have already been organized, and prospects for cooperation became firmer in October 1990, when two memoranda of cooperation were signed under the terms of the economic and commercial agreement of 1981. One of these deals with telecommunications, and applies to electronic components, data banks and databases, the computerization and networking of libraries, and ISDN.

Several activities are in progress with the Chinese Ministry of Posts and Telecommunications, Chinese regional authorities and universities concerning equipment testing, training centers, and computer networks.

Finally, information exchanges with Turkey have been strengthened recently.

—Relationships with Central and Eastern Europe
The development of telecommunications will be a major instrument for integrating Central and Eastern Europe into Greater Europe. Telecommunications will underpin economic reform in these countries; it is therefore imperative that the Community be in a position to respond to requests for assistance and cooperation.

The telecommunications sector in Central and Eastern European countries is acknowledged to be 20 to 25 years behind that of Western Europe due to chronic under-investment in the networks for many years. Institutional and regulatory reforms, as well as massive investments requiring a strong foreign component, are prominent among the necessary conditions for success in the development of infrastructures.

The commercial importance of the telecommunications market in these countries is widely recognized. The investment deals to be struck during the next few years will be the object of fierce competition, and the choice of technical standards will constitute a major issue.

At the "Arche Summit" in Paris in July 1989, it was agreed to provide support to Hungary and Poland. In July 1990, the group of 24 decided to expand assistance to Bulgaria, Czeckoslovakia, the German Democratic Republic (until the reunification of Germany), Roumania, and Yugoslavia. The European Bank for Reconstruction and Development (BERD), which will be operational in 1991, will contribute to the economic recovery of these countries. The European Investment Bank (EIB) is also preparing to intervene.

Measures for cooperation and assistance could take various forms. These include bilateral cooperative agreements which have been concluded recently, or are in the process of negotiation, between the Community and most of these countries. Other measures include coordinated aid from the Community and the Group of 24, general financial assistance from the BERD and the EIB, and "joint ventures" involving European industrial organizations (who could thus benefit from a market of 140 million inhabitants).

The PHARE program was allocated 300 million ECUs from the 1990 Community budget, and was aimed at assisting Poland and Hungary. A further 200 million ECUs has been set aside for other countries able to benefit from similar aid. Poland has planned activities in the telecommunications area under the terms of the PHARE program, and a project in rural telephony is already under way.

The Commission continues to examine the possibilities for action with governmental authorities, telecommunications operators and other concerned parties. Five lines of action are currently planned:

—exchange of information, standardization, and promotion of the mutual understanding of policies;
—integration into trans-European systems;
—business promotion and transfer of technologies;
—consultation and training
—activities aimed at promoting key developments in the telecommunications sector.

4

Towards
New Industrial Equilibria

4.1 ASSESSMENT

Since the mid-1980s, the long process of deregulation, the internationalization of markets, and the strong growth in demand made possible by technological innovation, have triggered an upheaval in the telecommunications sector.

In Western Europe as elsewhere, these movements have had two main effects. Firstly, industrial concentration among equipment producers pushing for de-partitioning of national markets. This is in order to allow European players to join forces and challenge the American and Japanese players in the great world markets, as well as to meet the demands of the single market.

The agreement signed between FIAT and CGE in 1990 shows that the era of national champions is drawing to a close, and that businesses have a common interest in closing ranks in the current economic war.

Figure 1 demonstrates the strong concentration in this sector, (weaker in Europe than in the United States and Japan because of the national markets). The ten leading companies control 54% of the total world telecommunications market.

This new economic situation is accompanied by a strong offensive in the European market by firms of other countries.

In market terms, Europe constitutes a prime target. This is primarily because it represents 30% of the world market, in second position behind the United States (50%).

Furthermore, it is a buoyant market: the aggregate total value of services and equipment in 1988 was of the order of 85 billion dollars, and it should at least double by the year 2000.

Finally, Europe is a new market in a number of new areas, where companies in other countries have sometimes acquired considerable

MARKET SHARE HELD	IN EUROPE	IN THE UNITED STATES	IN JAPAN	IN THE WORLD
>10%	ALCATEL-ALSTHOM SIEMENS	AT&T NORTHERN TELECOM	FUJITSU NEC	AT&T ALCATEL-ALSTHOM
5 - 10%	ERICSSON GEC	SIEMENS	HITACHI MATSUSHITA OKI	NEC NORTHERN TELECOM SIEMENS
2 - 5%	ASCOM AT&T BOSCH MATRA STET	ALCATEL-ALSTHOM GTE IBM NEC ROCKWELL	IWATSU MITSUBISHI TOSHIBA	ERICSSON FUJITSU GEC
COMPANIES				
The top 5	50%	55%	62%	41%
The top 10	62%	63%	74%	54%

Alphabetic listing in each group according to sales

Source EIC (Electronics in the world, November 1991)

Figure 1   Telecommunications hardware and equipment. Main companies (1990).

experience (for example, America with radio-telephone and cable TV, and Japan with facsimile).

If the three main market segments in this sector are considered (that is, equipment for operators, terminals, and services), the first results of this offensive are very significant. There are many convincing examples of this:

MARKET SHARE HELD	IN EUROPE	IN THE UNITED STATES	IN JAPAN	IN THE WORLD
>10%	ALCATEL-ALSTHOM ERICSSON GEC SIEMENS STET	AT&T NORTHERN TELECOM	FUJITSU HITACHI NEC OKI	AT&T ALCATEL-ALSTHOM NORTHERN TELECOM
5 - 10%		GTE/AG		ERICSSON SIEMENS
2 - 5%	AT&T MATRA	SIEMENS		FUJITSU GEC GTE NEC STET
COMPANIES				
The top 5	80%	96%	100%	60%
The top 10	89%	100%		75%

Alphabetic listing in each group according to sales

Source EIC (Electronics in the world, November 1991)

Figure 2   Public switching main companies (1990).

—with regard to public switching equipment (around $8.5 billion).

Figure 2 gives a good impression of the concentration, since the five top companies in the world hold 60% of the market, and the top ten hold 75%. The position of European industry remains still quite good.

However, AT&T, which only controlled 2.5% of the European market in 1985, controlled 13% in 1990, at the expense of European industry.

—with regard to telecommunications terminals (around $9.4 billion): European industry still retains dominance over a significant share of the market—mainly in top-range terminals. As in the United States, this dominance is partly due to the existence of a certification agreement.

On the other hand, in new and rapidly growing markets, such as radio and facsimile equipment (where Japanese industry holds 95% of the European market), European industry has in large measure lost its technological edge and suffers under a significant competitive handicap. Three factors favouring the development of the "grey" market are; the difficulty of mastering the distribution system, the absence of effective measures against fraud, and the developments in computers at the expense of dedicated terminals.

—In the field of telecommunications services (around $76.5 billion, of which $10 billion is in the non-telephone sector)

This area, which was until recent times non-competitive because on the monopolies of the national operators, is being progressively opened to competition, in particular the new rapidly growing services (radiotelephone, value added services, cable TV and distribution).

However, we are now seeing an important penetration of these markets by firms from outside countries. American operators (such as AT&T and the RBOCs) are present in the groups of consortia submitting bids for contracts in Europe. At the same time, the presence of other players, such as Motorola, should be noted in the service markets and the radio-mobile equipment market.

This American (in equipment and services) and Japanese (in terminals) penetration proceeds via agreements and participation arrangements with European firms.

An analysis of commercial and industrial agreements shows that the number of intra-European accords is roughly one third, while those between Europeans and non-Europeans comprise two thirds: (in 1989, 47% with American partners and 17% with Japanese firms). It should also be noted that the number of cooperative agreements with Japanese firms is continuously increasing. Even if the economic impact of these numerous agreements has not been well evaluated, it nevertheless translates into a strong internationalization of this sector.

Although it only depicts a brief period between July 1989 and January 1990, Figure 3 clearly demonstrates the magnitude of the phenomenon.

We are witnessing a veritable offensive on the European market by firms from outside countries with precisely targeted strategies. Due to their new competitive but regulated status, American operators and manufacturers consider it vital to invest in the European market. AT&T in particular has favoured the developing European countries (currently Spain and Italy, with Greece, Portugal and even the Eastern European countries to follow). The RBOCs, IBM, and Motorola for example, are positioning themselves in the market for new services in all the European countries.

Although proceeding more discretely, Japanese companies (following electronics and computing) would also like to get a foothold in the European telecommunications market. As they did with electronics, their offensive started with lower range products where European bids are less competitive. However, they have all the means needed to proceed progressively down the path to the more sophisticated products.

A few relevant examples can illustrate this.

● *The AT&T strategy in Europe*
The giant American AT&T is attempting to make its presence felt on the European market in both computing and telecommunications.

With regard to computing, its association with Olivetti since 1983 has not lived up to its expectations, which is apparently often the case when a telecommunications company tries to become involved in computing.

On the telecommunications side, the first AT&T operation in Europe was the creation of a joint company with Philips in 1984: APT (AT&T/Philips: 50/50). The arrival of Robert Allen at the head of AT&T, with a strategy of more direct intervention in Europe, saw AT&T's stake in APT go up to 60%, and a year later to 85%. The subsidiary then became AT&T NSI—AT&T Network Systems International. However, the operation was not a success and in the last quarter of 1990, Philips pulled out of this subsidiary.

In June 1987, AT&T opened negotiations in Spain to purchase Amper, a subsidiary of Telefonica, in return for a share in the operations of Marconi Espana, a company Alcatel was seeking to withdraw from. The section that interested AT&T specialized in the production of transmission equipment (hardware for network digitalisation) and employed 450 people. In October 1987, the agreement was finalized, and the company APT Espana was founded which was 51% owned by AT&T, and 49% by Amper.

The first signs of success were seen in 1989 when APT, along with Alcatel and Ericsson, was contracted to supply Spanish telecommunications with exchanges as part of their network modernisation plan.

Figure 3   Extra-European agreement, July 1989–January 1990.

Company		Company	Type of agreement
Italtel	<<>>	AT&T(US)	
Olivetti	<<>>	Sanyo/Mitsui (J)	Joint venture for manufacturing fax machines in Italy
Olivetti	<<>>	DEC(US)	DEC will carry out the distribution of Olivetti PCs in Europe
STET (I)	<<>>	GEIS(US)	Negotiations on a joint venture in value added networks
Olivetti (I)	>>>>	ISC Systems Corp (US)	Olivetti buys a "bank automation" company
Siemens (G)	<<>>	Unisys (US)	Cooperation in the development of a system for linking PABXs to computers (voice, data, network management)
Siemens (G)	<<>>	Matsushita (J)	Joint venture for passive components, resistances, valves
Bull (F)	<<>>	3Com Corp (US)	Bull will sell 3Com local networks in all markets under an OEM agreement
France Telecom (F)	<<>>	Apple (US)	Agreement on the development of ISDN connectivity (similar to the FT and ICL agreement)
Thomson (F)	<<<<	Motorola (US)	Thomson will carry out the marketing of certain Motorola microprocessors
Cable & Wireless	<<>>	US Sprint (US)	Agreement to install the first private transatlantic cable in competition with the national telecom companies
Racal (UK)	>>>>	Interlan (US)	Racal buys the American local network supplier
Aircall (UK)	<<<<	Bellsouth (US)	Bellsouth makes a bid for the 60% of Aircall that it does not already own
Cablelink (IR)	<<<<	Bellsouth (US)	Bellsouth makes a bid for the largest cable TV network operator in Ireland
Reuters (UK)	<<>>	Chicago Board of Options Exchange	Reuters will supply a global network for sale at "off-peak time"
STC (UK)	<<>>	NEC (J)	Cooperation in a joint venture for a cable between Europe and Japan
Amper (Sp)	<<>>	Motorola (US)	Amper (Telefonica Group) and Motorola in a joint venture to manufacture cellular switches and base stations in Spain

Company		Company	Type of agreement
Espanola de Microoordindores (Sp)	<<>>	Toshiba (J)	The Japanese group becomes a majority shareholder in the Spanish company in order to market its communication equipment
BT (UK)	<<>>	KDD (J)	KDD sets up a European subsidiary (Telehouse Int. Corp. of Europe) in which BT has 20% stake
MCI (US) + KDD (J)	>>>>	European PTTs	
BT (UK)	>>>>	Tymnet (US)	BT takes control of Tymnet (subsidiary of MacDonnel Douglas data communications)
AT&T (US)	>>>>	Istel (UK)	AT&T buys the systems integration company
IBM (US)	<<>>	Coditel (B) Sema (F)	RSNA joint service provision venture for businesses in the tourism sector
Ericsson (Sw)	>>>>	GE (US)	Cooperation in the manufacture of mobile communication equipment in Europe and Canada
AT&T (US)	<<>>	Olivetti (I)	Re-structuring of a previous agreement
FT (F)	<<>>	Toshiba (J) Matsushita (J) Murata (J)	Joint venture for developing high-speed fax machines in France

The chevrons (<<>>) indicate the direction of flow of investment or technology

Source: EEC/DG XIII

Figure 3   (*cont.*)

Intra-European agreements, July 1989–January 1990.

Company		Company	Type of agreement
STET (I)	<<>>	BT (UK)	Agreement in cross-border services
Teletra (I)	<<>>	Telenorma (G)	Joint venture in subscriber terminal equipment
SCS/Thomson (F)	<<>>	Thorn (UK)	Thorn sells Inmos (a subsidiary chip manufacturer) in return for 10% of SGS/Thomson shares
Sagem (F)	<<>>	Fiat (I)	Joint venture in navigation systems
Siemens (G) + GEC (UK)	>>>>	Plessey (UK)	Siemens and GEC buy Plessey after getting the approval of the European Commission and of R.U. (MMC)
Racal (UK)	>>>>	OTE (UK)	Racal undertake to assist OTE in launching an analog mobile telephone network
Philips (NL)	>>>>	PKI (G)	Philips pay 300 million DM for a minority share in its German subsidiary
Ericsson (Sw)	<<>>	EB (N)	Ericsson buys EB telecom division in exchange for signalling equipment
ACEC (B)	<<<<	Alcatel (F)	Alcatel becomes a partner in the SDT division (Space, Defence, Telecom)
France Telecom (F)	<<>>	ICL (UK)	Agreement on the development of ISDN connectivity (comparable to FT agreement with Apple)
Alcatel (F)	<<>>	Siemens (G)	Technical cooperation on ISDN switching standards
Philips (NL)	<<<<	Thomson (F)	Philips sells 49% of its defence interests to Thomson
Ericsson (S)	<<>>	Bang & Olufsen (DK)	Joint venture in small switches
Matra (F)	<<>>	GEC (UK)	Amalgamation of their aerospace activities (Matra-Marconi Spatial)
Alcatel (F)	<<>>	PTT Finland (SF)	Cooperation in the development of a metropolitan network
Nokia (SF)	>>>>	NKF (NL)	Nokia buys a share of Dutch cables

The chevrons (<<>>) indicate the direction of flow of investment or technology

Source: EEC/DG XIII

Beyond this national success, the Spanish unit was intended to act as a springboard to launch AT&T in the Mediterranean market. This was confirmed during the summer of 1989 by ITALTEL's the success in Italy with a technological, commercial and financial agreement.

After it had received authorization from Judge Greene to enter the value added services market on August 24, 1989, AT&T also decided to acquire the British computer services company ISTEL.

At the end of 1990, Italtel and Telefonica joined AT&T NSI.

There is no doubt that AT&T will continue its efforts to penetrate the European market.

● *The arrival of the RBHCs in Europe*
The European policy of the RBHCs is reflected perfectly in their overall strategic plans for diversification into unregulated activities (cable TV, cellular networks and computer services) and in their desire to become international giants at the service of their clientele—the multinational American firms.

The strategies adopted are diverse, from maintenance to computer services, and include network administration. Viewed geographically, the United Kingdom, Spain, and France are their initially preferred zones, moving towards Eastern Europe, as the activities of Nynex and US West in Hungary, for example, have illustrated.

● *The IBM presence in Europe*
IBM has tackled the telecommunications sector via various aspects of data transmission, including equipment (modems and multiplexers), value added networks and services.

The IBM posture in Europe is built on "strategic" contracts in step with the liberalization of the market.

1984: A project with BT in the value added networks field, which was aborted due to the British authorities.

1987: an association, also in the value added networks domain, with Paribas and Sema-Metra in France; Telefonica in Spain; FIAT in Italy.

1987: an association in the field of intelligent networks with Siemens in Germany, and Ericsson in Sweden.

Finally, also in 1987, an association in Denmark with KTA in the field of electronic messaging.

1989: in Belgium, an association with Coditel and Semagroup Belgium to create a value added network provider company for businesses in the tourism sector.

At the end of 1990: IBM pursued its advance by concluding an agreement with British Telecom.

More events are coming....

GERMANY

AMERITECH - Radio-communications: Partnership in a consortium for a second cellular network. Acquisition of Germany Yellow Pages publishing company in 1990

Bell Atlantic - Computer services: Owns SORBUS, a Computer Services Company (1987), with subsidiaries across Europe—Radio-communications: Partnership in a consortium for a second cellular network

BELL SOUTH COMPANIES INTERNATIONAL (BSI) - Partnership in a consortium for a second German cellular network (Sept. 1989 associated with BMW, Veba, Racal)

NYNEX - Partnership in a consortium for a second cellular network

PACIFIC TELESIS - Partnership in a consortium (Mannesman) for a second cellular network (26%)

USWEST - Partnership in a consortium for a second cellular network

BELGIUM

AMERITECH - Bought ADR (Applied Data Research) in 1986, then transferred to Computer Associates International Inc. in 1988. American SSII installed in Europe (UK, Belgium, France)

Bell Atlantic - Installed in Brussels in 1989

SPAIN

AMERITECH - Partnership with Telefonica Nacional de Espana

Bell Atlantic - Installation of a network management system for Telefonica (1987)

NYNEX - Creation of a joint venture for network operations in Gibraltar (1989)

PACIFIC TELESIS - R&D centre with Telefonica

FRANCE

AMERITECH - (1990) Consortium with France Telecom to construct and operate Poland's national cellular telecommunications system (49%)

AMERITECH - Bought ADR (Applied Data Research) in 1986, then transferred to Computer Associates Internatonal Inc. in 1988. American SSII installed in Europe (UK, Belgium, France)

BELL SOUTH COMPANIES - Installation in Metz in 1986; Creation of Datech in 1987
- Cellular venture Bell South (4%), Generale des Eaux, Nokia, Racal

USWEST - Owns 10% of Lyonnaise Communications

NYNEX - Computer services; purchase of BIS (Business Intelligence Services) in December 1986

Bell Atlantic - Eurotechnica France 1987, purchase of computing activities of Bell Canada Enterprises

HOLLAND

Bell Atlantic - Traffic monitoring system for the Dutch PTT (end of 1988)

AMERITECH - Partnership with PTT telecom BV

ITALY

AMERITECH - (1990) 3 years agreement with Italcable in international projects

Bell Atlantic - Agreement on network technology with Italtel: Eurotech Italia 1987 (purchase of computing activities of Bell Canada Enterprises)

BELL SOUTH COMPANIES - Contract with SIP (end of 1987)

UK

AMERITECH - Bought ADR (Applied Data Research) in 1986, then transferred to Computer Associates International Inc. in 1988. American SSII installed in Europe (UK, Belgium, France): British Voice Messaging

BELL SOUTH COMPANIES - In 1986, 40% stake in Aircall (Radio-communications company), and 100% in 1989: Purchase of the radio-telephone branch of National Radiophone (Birmingham 1986): Partnership in a PCN consortium: Acquisition of Dataserve (Maintenance and leasing company for IBM products) SSII in 1986: Subsidiary of Bell South International in 1985

NYNEX - 1990 awarded 11 cable franchises Network management system for BT (1987)

PACIFIC TELESIS - Interests in three cable TV companies (East London Telecommunications Ltd, Peterborough Cablevision, and Norwich Cablevision): Partnership in a PCN consortium: In consortium (20%) with British Aerospace PLC

USWEST - 25% of Cable Corp., 34% of Cable London
- Participation in a consortium for PCN
- 30% of West Country Cable

Southwestern Bell Corporation - Cable televison

Bell Atlantic - Consultancy

Source: Ministry of European Affairs (France 1990)

Figure 4   Positioning of the American RBOCs in European telecommunications.

● *The assertion of Motorola on the European market*

Motorola began in Europe in fifth position on the components market (after Philips, SGS, Thomson and Texas Instruments); its ambition is to take second position on the European market by 1992.

In another area, Motorola is currently taking up positions throughout the rapidly expanding European radio-telephone market. The objective of Motorola, (tied with Philips at 13% of the European radiocommunications market) is to quadruple its sales. To this end, in 1986, the company bought Storno (a subsidiary of General Electric) which has branches in Denmark, Great Britain and France.

It is also a partner with Mercury in the Cellnet network in Great Britain.

In July 1989, Motorola also launched a project on the future British mobile PCN telephone system in partnership with Mercury.

The activities that have just been described have been reported in the press; they represent only those that have been observed, and are by no means the only ones. Different players, after the first attempts undertaken to test the water, are managing to progressively define viable strategies.

Other American companies such as GTE, Rockwell, Northern Telecom, Corning Glass are now moving into the European market. Figure 5 shows some of the principal penetrations of American companies in Europe.

● *The Japanese approach is very different*

The entry of the Japanese companies into the telecommunications market is being achieved via the mass production and electronics sectors.

These companies have both an industrial strategy and a multi-disciplinary approach: telecommunications forms part of a series of other activities including computing, mass-market electronics, and semi-conductors.

As regards telecommunications products in the strict sense, the Japanese are present in the area of terminals (fax, and telephones) above all in Great Britain (NEC, Ricoh, and Matsushita), and France (Canon, Ricoh, Sharp, and Mitsubishi), but also in Spain and Italy.

In the semiconductors field (particularly in memory production) two main manfacturing zones exist: in Germany (Mitsubishi, Matsushita, Hitachi, Toshiba), and in Great Britain (NEC, Fujitsu, Omron).

Japanese investments in Europe are already sizeable, and they will further increase within the context of the 1992 market.

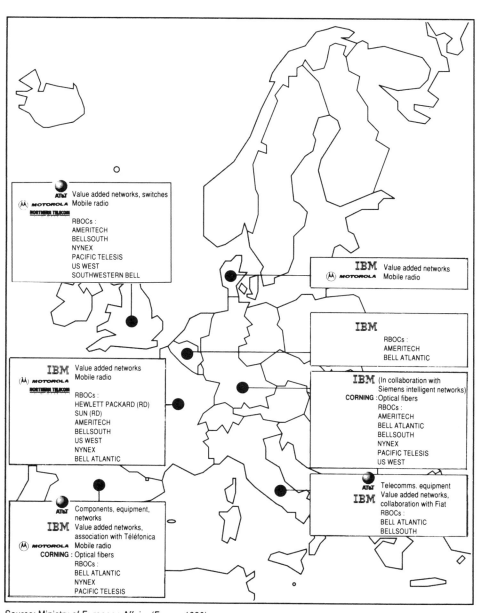

Source: Ministry of European Affairs (France 1990)

Figure 5 American firms in European telecommunications.

Source: Ministry of European Affairs (France 1990)

Figure 6   Japanese industrial and commercial presence in Europe.

LEGEND

P	Photocopy	TV	Television
T	Fax	Ma	Video recorders
EGP	Mass-market electronics	tel	telephones
R	Radio-telephone	fc	Subsidiary marketing
I	Computing	TM	Modems and terminals
SC	Semiconductors	Tel	Telecommunications

IRELAND
FUJITSU :   I (84), SC (80)
NEC :   SC (74)
SAWAFUJA :   R

GREAT BRITAIN
CANON :   Research center
FUJITSU :   Europe Ltd, SC (89),
                buys ICL (90) Tel
HITACHI :   Europe Ltd, EGP (84), TV, Ma
KDD :   1 subsidiary, I (research
                laboratory 87)
MATSUSHITA :   R (88), tel (76), T (89), EGP
MITSUBISHI :   Ma, TV
NEC :   UK Ltd, R (88), SC (83, 86)
                T (89)
OKI :   Europe Ltd
RICOH :   P (84), T (89)
SANYO :   TV, Hi-Fi
SHARP: P, EGP
SONY :   TV
TOSHIBA :   TV

SPAIN
CANON :   fc
FUJITSU-TELEFONICA : TM (86)
HITACHI :   fc
KDD :   Agency, representatives office
NEC :   fc
MATSUSHITA :   EGP
MITSUBISHI :   fc
RICOH :   fc
SANYO :   TV, Hi-Fi
SONY :   EGP
TOSHIBA :   fc (89)

DENMARK
HITACHI I :   fc
SONY :   Scandanavia fc

NETHERLANDS
CANON :   fc
FUJITSH :   fc
HITACHI :   fc
MATSUSHITA :   fc
MITSUBISHI :   fc (72)
NEC :   fc
RICOH :   fc
TOSHIBA :   fc
SONY :   fc

BELGIUM
HITACHI :   fc
MATSUSHITA :   Piles (70)
SONY :   fc
KDD :   Agency

FRANCE
CANON :   P (83), T (86)
HITACHI :   fc
KDD :   (Agency)
MATSUSHITA :   EGP
MITSUBISHI :   R (88)
NEC :   fc (88)
OKI :   (Agency)
RICOH :   P (88), T (89)
SHARP :   P (88), T (89)
SONY :   Cassettes (80   82)
                Compact disc reader (85)
TOSHIBA :   P (Toshiba Rhône Poulenc 86)
                P (87), T (Toshiba - Telic 89)

GERMANY
CANON :   P (72)
FUJITSU :   2 fc
HITACHI :   Europe GmbH
                EGP (83), SC (80)
KDD :   Representatives office
KONICA :   P (87)
MATSUSHITA :   P (87), SC (89), EGP (83)
MITSUBISHI :   Electric Europe GmbH
                SC (in construction)
NEC :   Electronics Europe
OKI :   Electronics Europe
RICOH: 1 fc
SANYO :   Ma
SHARP :   1 fc
SONY :   Europe GmbH, TV, Hi-Fi
TOSHIBA :   SC (84, 89), EGP (86),
                I (89)

GREECE
MATSUSHITA :   1 agency
SHARP :   1 subsidiary

ITALY
CANON-OLIVETTI : P
FUJITSU :   2 fc
KDD :   1 representatives office
NEC :   Research center SC (88)
MATSUSHITA :   1 fc
SANYO-OLIVETTI. MITSUI : T
SONY :   Cassettes (87)
TOSHIBA :   1 fc

PORTUGAL
MATSUSHITA :   Agency
SONY :   fc

Japanese companies can be classified into three categories:

—companies in which the telecommunications activity is important: NEC, Fujitsu, Hitachi, OKI;
—large companies in which telecommunications is far from the most important activity: Matsushita, Sony, Toshiba, Mitsubishi;
—small scale companies specialising in office equipment and firmly established in Europe: Canon, Ricoh, Sharp, Sanyo.

Figure 6 gives a good overall picture of the penetration of these companies in Europe at the beginnning of 1990.

4.2   IN CONCLUSION

If the overall commercial balance in the Community's telecommunications sector is currently in quasi-equilibrium, it should be noted that there is a significant deficit with the United States ($680 million) and Japan ($1360 million) in the equipment and terminals market. These figures, however, do not represent a true economic overview because they do not integrate the balance of services with the balance of capital. It should also be noted that these deficits appear very high when compared to the level of trade with these countries, which is still very modest, representing only 3% of the Community equipment market.

This assessment shows the clear positions on the Community market taken by firms from outside countries. This trend will continue, or even accelerate.

This phenomenon can be explained in several ways:

—the liberalization of services in progress at the Community level, which will probably be combined into an agreement on services within the framework of the GATT talks (Uruguay Round). In the negotiations on services, the United States has made telecommunications one of its priorities. New telecommunications services, henceforth in the competitive arena, are now growing strongly; trans-national services could be an especially favoured target for operators from other countries, because they are already experienced in the supply of these services (e.g. IBM in value added services, Motorola in mobile telephony...)
—The opening of the public markets will allow easier access to firms from other countries; the service providers will be able to overcome the constraints of industrial policy imposed by the States.

—The implementation of the internal market will result in the free circulation of telecommunications equipment and the unimpeded establishment of service providers in the Community. In the field of terminals there is a risk that, if care is not taken, certification systems will sink to the lowest common denominator: it will be important to ensure that the Community's certification system operates with a satisfactory level of quality and uniformity.

In a general sense, these changes will benefit both operators and the manufacturers of other countries, whose access to the Community market will be facilitated.

In the face of this situation, the European players have only achieved a limited presence in the American market, and are absent in the Japanese market. Furthermore, they have not adopted a resolutely aggressive stance in these markets, as the following brief analysis illustrates:

The American market, currently the most important, is in theory open to European operators. However, in reality the market is much less attractive.

The European penetration in the services sector is still in a virtually embryonic stage. The entry of European operators, even by way of partnerships, appears to be extremely difficult, partly due, of course, to the competition, but also to regulations which are extremely complex, rapidly changing, and make use of different standards. The price of entry is high, and success requires a spirited long-term strategy. It is a means of access for European operators however, notably in the newer market segments (data transmission, mobile communications and value added services), where they have real technological strengths to work from.

The American network equipment market has only been penetrated by two European companies: Siemens, (switching), and Ericsson (mobile telephony). They have, however, enjoyed great success. Siemens, in allying itself with GPT at the end of 1990, has become the number three among the major suppliers of network equipment, after AT&T and Northern Telecom. Ericsson (thanks to a contract obtained in October 1990 for the supply of mobile phones to the American group McCaw Cellular Communication Inc. worth 1.8 billion dollars over ten years) has overtaken its competitors Motorola and AT&T, and controls more than 30% of the North American market in mobile phones. The modification of the tariff system negociated by the Federal authorities (Cap price) should imply a need by the BOCs to renew their infrastructures. This could present a new opportunity for European industry.

As in the terminals sector, the weak competitive position of European offerings compared with those from Japan and South East Asia makes it difficult for them to make any inroads in this sector.

With regard to the Japanese market, a significant European break-through seems unlikely, although the deregulation process is in under way. However, it should be possible to find openings through targeted opportunities. It should be underlined that American pressure to reduce its general trade deficit makes the possibilities of significant advances into the American and Japanese markets more difficult. The consequences of this pressure are obvious:

—an increase in disguised American protectionism,
—American industry is to be given priority in any opening in the Japanese market (e.g. Motorola),
—a movement to relocate Japanese industry to the United States, with re-exportation to Europe.

There is no miraculous solution that will reverse this trend in the short term: however, returning to a protectionist system is virtually impossible on the political level, and inappropriate on the economic level. It is essential that the European industrialists adopt a much more aggressive approach; they should seize all opportunities that present themselves in the American and Japanese markets, and the markets of other countries. At the same time, they must continue to assert their presence on the European market. To accomplish this, the European players should seek out growth in all its forms: internal growth, external growth (buying-in to joint ventures, stock market acquisitions), and partnerships.

The major operators should devote substantial investments to these objectives, even if their short term viability is not assured. To this end, consortia should be set up, with the European players deliberately acting in a coordinated manner without refusing, how-ever, agreements with American firms, because such agreements are often involved in the conditions stipulated in requests to tender for contracts.

The strategy of Cap-Sogeti can be cited as an example of tactics used in this field. The company gained a strong position on the American market in order to counter an offensive from American firms in the European services market. To achieve this, the company formed joint ventures, and made strategic purchases and cooperation agreements.

This effort in the international arena implies a strengthening of European bids for contracts. This kind of strategy, aggressively confronting American and Japanese firms, requires truly strong European players: a high level of investment and R&D, and reinforcing cooperative arrangements, particularly between operators. This in turn implies greater autonomy for management and an adequate financial capacity for the public network operators. Strengthening of the European players no longer represents a risk to consumers, because the environment in this sector will become highly competitive.

At the internal level, the Community should play its full supporting role in the face of the threat facing the European players.

The Community should continue to create an environment which is favorable to European firms by an active policy:

—incentives for R&D (the program framework should strongly support telecommunications development);
—standardization (the creation of ETSI is a significant first step, but substantial efforts to improve the efficiency of the European standards system remain to be made);
—structural Community funding devoted to financing the telecommunications infrastructure (such as the STAR program), should be used, within the rules concerning the awarding of public works contracts.
—large high-profile projects: the Community should not substitute itself for the current players (it has neither the means nor the competence, and the risk of bureaucratization would be significant), but rather it should devolve its initiatives after having defined the priorities. A European label would of itself add dynamism to such projects, for which a cooperative approach would be encouraged. The "trans-European networks" initiative adopted at the Strasbourg Summit could provide an appropriate framework for this purpose.

Depending on the way that the Commission applies article 85, competition could be exercised without abusing dominant positions, while protecting consumers as the Treaty specifies, or, by forbidding useful alliances, it could lead to a patchwork of multiple networks in Europe.

If one takes the North American example, Canada and the USA have agreements and understandings both at the regulatory authority level, and between American companies.

Futhermore, cooperative agreements can not only be viewed from the perspective of the rules of competition. Their analysis should also take into consideration the industrial and economic interests of the Community.

On the external level, the Community should identify existing or potential blockages in outside markets, in order to insist on their removal during future negotiations (multilateral or bilateral).

Opening the European market to competitive services and achieving the desired conditions of access to the network cannot be accomplished without effective reciprocity: the concept of dealing with this at a national level alone is insufficient, because it does not prevent the discriminations practised by locally based authorities.

As regards the Japanese, the proper opening of the telecommunications services sector to the benefit of European interests should be negociated within a bilateral framework with the Community, with a view towards redressing the imbalance of trade between the Community and Japan.

Finally, the Community should reinforce its anti-dumping procedures (particularly with regard to terminals) by a much more rapid examination of dossiers, and even the implementation of a proper procedure emergency ruling.

4.3   EPILOGUE

This study of the recrystallization of world telecommunications stops at a point when nothing is settled. The transformation follows its turbulent path without giving any hint of the ultimate form that, one day, a calmer stream will lead it to. The original source of this structural metamorphosis at the level of State monopolies, industries, and of demand, has unquestionably been the extraordinary evolution of technology. The birth and prodigious growth in microelectronics, combined with the general implementation of information digitization, have brought enormous upheavals in products and services offered to customers. Consequently, a fundamental re-examination of telecommunications was inevitable. Started in the United States, this renewal spread to the rest of the world and, in particular, the entire European Community, which has recognised how to lay the foundation for a significant response.

Technology is following its course, but one notices fewer new fundamental breakthroughs being announced. We are now witnessing the challenges of mastering microelectronics and digital techniques. This phenomenon continues to be of considerable importance, but it no longer seems to contain any surprises of the first order: the margins of technical uncertainty are now decreasing, except perhaps with regard to the reaction of users when faced with new services, and the effects of this reaction on techniques.

From another point of view, the political environment which prevailed during the decade 1979–1988 has been abruptly changed. The countries of the European Community must integrate major modifications into their strategies, affecting their relationships with Eastern Europe and the Soviet Union. The future of our continent will be profoundly influenced by the quality of the trading methods, resulting from the deployment and transparency of the telecommunications systems. In large measure, Community policies concerning investment, Research and Development, and standardization will have to take into account this opening-up of the geographic zone, and the specific needs the Central and Eastern European countries have in their difficult process of adapting to a market economy.

In addition, apart from the East, the success of the initiatives of the Commission and the proximity of the opening of the "single market" have had the effect of inspiring calls for a "rapprochement" from countries that had not previously sought such close links. This applies particularly to the interest shown by certain Scandinavian States, Austria, Turkey, and Morocco.

The wish sometimes expressed for a political cohesion poses vast problems, but the development of organic links in the fields of scientific, technical and industrial cooperation would be a positive immediate response to these aspirations. Telecommunications is clearly at the forefront of those specialties in which such cooperation could flourish.

The economic environment has also been profoundly modified. The inevitable uncertainties accompanying the inauguration of the "single market" have been compounded by a reversal of the current regulatory situation. Among traditional telecommunication equipment and services suppliers, and industries that could be called "new recruits", the growth in revenues and the easing of profit margins, herald an era of very open competition. From 1989 in America, but above all in Europe, the electronics and computer markets have become less buoyant, competition from Japan and South East Asia has become more incisive, and the margins of many industries specializing in domestic equipment have been reduced or become negative. This situation could significantly limit their capacity to intervene in the telecommunications arena. Some companies have abandoned these objectives and have given up their specialized departments. Others, on the contrary, are tempted to move towards this promising new market, accepting the unusual risks of re-conversion, alliances, or joint venture activities.

In summary, it can be said that the factors of uncertainty and instability are more present than ever, but that they have shifted from the purely technical towards national political initiatives and business strategies. In this situation, the European Community is a central point for the worldwide redistribution of telecommunications activities. This is because of the importance of its internal and associated markets, and its being regarded by the large multinational companies as being "open", or even "available". In contrast, the American economic zones (and even more so, the Japanese) have effective means of controlling their markets.

A warning must be added to all of these questions, arising from the Persian Gulf crisis. The behaviour of Iraq is a reminder of the necessity for making Europe less dependent on Middle East fossil fuel sources. This objective essentially translates into a need to reduce energy consumption in the medium and long term. The importance of an information technology policy, featuring the use of telecommunications as a substitute for the transportation of people is all the more imperative in the wake of the Gulf War, started by Iraq's invasion of Kuwait on August 2 1990. If a transformation of energy needs cannot be attained, Western economies will remain vulnerable. Therefore, policy must be corrected by appropriate measures, among which the policy for teleçommunications, information technology, the information industries, and industrial innovations will play an important role.

Of course, the telecommunications revolution will remain dominated by the development of services linked to inter-communication of "intelligent images", of "organized data", and by interactive information processing. These new so-called "value added services"—in comparison to the simple transmission of coded information—give us a glimpse into the partly unpredictable future. The relative failures of professional teletex in Germany and the United Kingdom, compared to the prodigious success of Minitel in France were surprises. Only truly large scale experience can yield accurate answers. In the same vein, the current explosive growth of facsimile was not anticipated during its years of slow progress. What will become of mobile phones (which are effectively cordless terminals), for which multiple applications are proposed? How will all this technological progress link with the inevitable consequences flowing from the achievements in high-definition television (HDTV)? It can be seen that policies will need to closely follow the demands and tastes of users. The reactions of consumers will need to be stimulated by bold initiatives anticipating needs that have not yet been expressed, nor even yet experienced. These are important and difficult objectives if one takes into account the need to attain maximum efficiency under the unavoidable constraints of budgetary rigour. These initiatives will nevertheless be necessary. Europe cannot limit itself to following the Americans or the Japanese while leaving to its competitors the advantages of conquering the first market; that is, the benefits from the first generation of mass-market products which ensure high profit margins and provide the means to finance new progress. The Europe of the Single Market will also need to respond in a climate of total freedom of industrial intellectual property, patents, and technology transfer, to demands from the nations on its Southern and Eastern borders, who expect it to lead them down the paths to their own development.

This telecommunications adventure is only just beginning. More than ever, it confirms that the scale of the problems justifies Community solutions. The battles of research and development, total mass-production costs, installation of sophisticated equipment and multiplication of subscribers outlets, control of norms, standards, and approvals procedures, and diversification of new services can only attain a global dimension by starting from a large enough demographic base. This is precisely what is offered by the Community dimension, and the existence of a framework of strategies clear enough to be effective, but flexible enough to leave each member country the freedom to adapt to its own particular situation.

List of Acronyms

AIM:	Advanced Informatics in Medicine
AMERITECH:	American Information Technologies
AMIS:	Agricultural Market Intelligence System
AOS:	Alternative Operator Services Company
ASST:	Azienda di Stato per i Servizi Telefonici
AT&T:	American Telephone and Telegraph
AT&T CIS:	American Telephone and Telegraph Communications and Information Systems
BT:	British Telecom
CADDIA:	Cooperation in Automation of Data and Documentation for Imports/Exports and the management of the financial control of the agricultural market
CCIR:	International Consultative Committee for Radiocommunications
CCITT:	International Consultative Committee for Telephones and Telegraphs
CEI:	Comparability Efficient Interconnection
CEN:	European Standards Committee
CENELEC:	European Electrotechnical Standards Committee
CEPT:	European Posts and Telecommunications Conference
CERN:	Centre d'Etudes et de Recherches Nucléaires
CNTE:	Compania Nacional Telefonica de Espana

CPE:	Customer Premises Equipment
CPRM:	Compania Portuguesa Radio Marconi
CTT:	Correios Telecommincações de Portugal
DECT:	Digital European Cordless Telephone
DELTA:	Development of European Learning through Technological Advance
DGT:	Direction Générale des Télécommunications
DIANE:	Direct Information Access Network for Europe
DOJ:	Department of Justice
DRIVE:	Dedicated Road Infrastructure for Vehicles in Europe
EC:	European Community
EDI:	Electronic Digital Interchange
EDIFACT:	Electronic Data Interchange for Administration, Commerce and Transport
EDS:	Electronic Data Systems
EFTA:	European Free Trade Association
EIB:	European Investment Bank
EMS:	European Monetary System
ETSI:	European Telecommunications Standards Institute
FCC:	Federal Communication Commission
FIS:	Fast Information System
GAF:	Group Analysis and Forecasting
GATT:	General Agreement on Tariffs and Trade
GOT:	Greek Office of Telecommunications
GSM:	Special mobile Communications Systems
GTE:	General Telephones and Electronics
HDTV:	High-definition Television
IBC:	Integrated Broadband Services

ICCP:	Information, Computing and Communications Policy Committee
IDES:	Interactive Data Entry System
INSIS:	Inter-institutional integrated Services Information and Documentation
ISDN:	Integrated Services Digital Network
ISO:	International Standards Organisation
IT:	Information Technologies
ITT:	International Telephone and Telegraph
ITU:	International Telecommunications Union
LAN:	Local Area Network
LATA:	Local Access and Transport Area
LOT:	Law for the Organisation of Telecommunications
MAFF:	Market Access Fact Finding
MFJ:	Modified Final Judgement
NEC:	Nippon Electric Company
NTIA:	National Telecommunication and Information Administration
OCC:	Other Common Carrier
OECD:	Organisation for Economic Cooperation and Development
OFTEL:	Office of Telecommunications (GB)
ONA:	Open Network Architecture
ONP:	Open Network Provision
OSI:	Open Systems Interconnection
PABX:	Private Automatic Branch Exchange
PUC:	Public Utility Commission
RACE:	Research and Development in Advanced Communication Technology for Europe
RBC:	Regional Bell Clients
RBHC:	Regional Bell Holding Company

RBOC:	Regional Bell Operating Company
R&D:	Research and Development
RHC:	Regional Holding Company
ROES:	Receive Only Earth Station
RTT:	Régie des Télégraphes et des Téléphones
SCC:	Specialised Common Carrier
SCENT:	System Customs Enforcement Network
SGP:	System of Generalised Preferences
SIP:	Societa Italiana per l'Escercizio Telefonico
SOGITS:	Senior Official Group for Information Technologies Standards
SOGTS:	Senior Official Group for Telecommunications
SPAG:	Standards Promotion and Application Group
STAR:	Special Telecommunications Action for Regional Development
STID:	Scientific and Technical Information and Documentation
TARIC:	Tarif Intégré Communautaire
TEDIS:	Trade Electronic Interchange Systems
TLP:	Telefones de Lisboa e Porto
TRADE ACT:	Trade and Competitiveness Act
USTR:	United States Representative
VAN:	Value Added Network
VAS:	Value Added Service
VSAT:	Very Small Aperture Terminal
WATTC:	World Administrative Telephones and Telegraphs Conference

Index

Advanced Informatics in Medecine
    (AIM) 110
Allnet 29
Alternative Operator Services
    companies (AOS) 33
Ameritech 24
APT 149
Arche summit 143
ASST (Aziendo di Stato per i Servizi
    Telefonici) 79
AT&T 1, 4, 7, 27, 149, 153

Belgium 75, 125
Bell, Graham 2
Bell Atlantic 24
Bell Labs 9
Bell Operating Companies (BOCs)
    7, 18, 32
Bell South 24
BellCore 25
British Telecom (BT) 60, 61, 77,
    133
British Telecommunication Act 77

Cable Television (CATV) 38, 41
CADDIA 51, 56, 57
Canada 163
Cap-Sogeti 162
CE monitoring 112
Cellular radio communications 43,
    103
Communication Act (USA 1934) 4
Community Information
    Technology policy 47
Compania Nacional Telefonica de
    Espana (CNTE) 76
Comparably Efficient
    Interconnection (CEI) 70

Competition (Europe) 60
Competition (USA) 13, 31
Computer Inquiry I, 17
Computer Inquiry II 17, 18
Computer Inquiry III 31, 36
COMSAT 11
Consent Decree 5, 19,
Continental Telecom 11
Cordless phones 106
Council of Ministers 53, 88, 116

Dedicated Road Infrastructure for
    Vehicle Safety in Europe
    (DRIVE) 110
Delors Commission 87
Denmark 75, 126
Department of Justice (DOJ) 5, 16
Deregulation (USA) 17, 29
Deutsche Bundespost 63, 81, 127
Developing countries, relationships
    with 141
Development of European Learning
    by Technological Advance
    (DELTA) 110
Digital European Cordless
    Telephone (DECT) 106, 107
Direct Information Access Network
    for Europe (DIANE) 55
Direction Générale des
    Télécommunications (DGT,
    France) 76
DOMSATs 13

Eastern European countries 142
Electronic Data Interchange (EDI) 59
Electronic Data Interchange For
    Administration Commerce and
    Transport (EDIFACT) 58, 59

Electronic publishing 35
EN standards 70
Equipment market (USA) 39, 42
Ericsson 15, 151
ESPRIT program 47, 52
ETSI 70, 102
EUROFARM 58
EURONET 54, 55
European Bank for Reconstruction
    and Development (BERD) 143
European Commission 49, 55, 68,
    118, 137
European Court of Justice 60
European Economic Community
    (EEC) 89, 98
European Economic Zone 141
European Electrotechnical
    Standards Committee (EESC)
    66
European equipment and services
    markets 84, 115
European Free Trade Association
    (EFTA) 59, 68, 141
European Funds for Regional
    Development (FEDER) 136
European Investment Bank (EIB)
    143
European Parliament 118
European Posts and
    Telecommunications
    Conference (CEPT) 53, 70, 92
European Single Market 87, 167
European Standards Committee
    (CEN) 66, 68
European Telecommunications
    Zone 89, 100
Execunet 7

Fast Information Interchange (FIS)
    59
Federal Communication
    Commission (FCC) 4, 5, 6, 19,
    31, 34
Federal Court of Appeals 36
Federal Republic of Germany 78,
    127
France 76, 126
France Télécom 126

General Agreement on Tariffs and
    Trade (GATT) 91, 95, 98, 138

GPT 161
Great Britain 77, 133, 134
Greece 78
Green Paper 94, 96, 100
GSM (Special Mobile
    Communications Systems) 104
GTE-Sprint 85
GTE 11, 25

High definition television (HDTV)
    166
Hush-a-phone 5

IBM 30, 149, 153
Industry, role of 70
Information, Computing and
    Communications Policy
    committee (ICCP) 139
Information Technology 48, 90, 94
Integrated Broadband
    Communications (IBC) 107,
    109
Integrated Systems Digital Network
    (ISDN) 66
Inter-institutional Integrated
    Services Information and
    Documentation (INSIS) 56
International Consultative
    Committee for Radio
    communications (CCIR) 90
International Consultative
    Committee for Telephones and
    Telegraphs (CCITT) 66, 91,
    102
International Standards
    Organisation (ISO) 66, 70
International Telecommunications
    Union (ITU) 91, 139
International Telephone and
    Telegraph (ITT) 83
Ireland 78, 128
ISDN 90, 92, 102, 135
Italy 78

Japan 140, 156, 160, 167
Judge Harold Greene 19, 26, 35

Legal action, USA 5
Local Access and Transport Areas
    (LATAs) 25

Lomé Convention countries (ACP) 142
Luxembourg 78, 130

Market access fact finding (MAFF) enquiries 87
Mann-Elkins Act 3
MCI 6, 29, 85
Mercury Communications 78, 133
Mobile telephony 40, 104
Modified Final Judgement (MFJ) 19, 25, 37
Motorola 156

National Telecommunications and Information Administration (NTIA) 29
NEC 15
Netherlands 7, 130
Northern Telecom 45
NTT 138, 139

Objective '92, 87
OECD 139
Office of Telecommunications (OFTEL, UK) 78, 134
Omnibus Trade Act 87, 140
Open Network Architecture (ONA) 32
Open Network Provision (ONP) 98, 118–125
Open Systems Interconnection (OSI) 70
Optical fibers 43
Other common carriers (OCCs) 7, 13, 27

Pacific Telesis 24
Packet switching 117
Payphones 33
PHARE 143
Plessey 15
Portugal 78, 131
Price cap (AT&T) 34, 161
Private Automatic Branch Exchange (PABX) 43
Public Transportation Act 3

RACE program 94, 109, 110
Radio communications 30

Radio paging 40, 104
Receive Only Earth Stations 98
Regional Bell Holding Companies (RBHCs) 21, 26, 153
Regional Bell Operating Companies (RBOCs) 24, 26, 35, 148
Research and Development programs (Europe) 52, 109

Satellite communications 30, 42, 43
SCENT 58
Sherman Act 5
Siemens 15
Single Market Act 73
SIP 79
Social issues 134, 136
SOGITS 68
SOGT 91, 98
South Western Bell 24
Spain 75, 132, 149
Standardization 64, 111, 115
Standards Promotion and Application Group (SPAG) 71, 73
STAR program 94, 99, 135

Tarif Intégré Communautaire (TARIC) 53
Tariffs 85
Telecommunications services markets 84, 115
Telefonica 133
Telephone companies, by country 75
Telepoint 106, 107
Teletel 107
Trade and Competitiveness Act 45
Trade Electronic Interchange Systems (TEDIS) 56, 59
TRANSPAC network 55
Treaty of Rome 97

UKCT2 standard 106
United States Trade Representative (USTR) 87, 98
United Telecommunications 11
Uruguay round 138
US Congress 4, 5
US Court of Appeals 35, 36, 38
US Sprint 28
US West 24

Vail, Theodore 2
Value added networks 42
Value added services 95, 97

Western Electric 9, 13
Williams Communications 29

Willis Graham Act 4
Wireless communications 29
World Administrative Telephones
    and Telegraphs Conference
    (WATTC) 139
World markets 145